Advances in Food and Beverage Labelling

Related titles

Foods, Nutrients and Food Ingredients with Authorised EU Health Claims
(ISBN 978-0-85709-842-9)

Dietary Supplements: Safety, Efficacy and Quality
(ISBN 978-1-78242-076-7)

Managing and Preventing Obesity: Behavioural Factors and Dietary Interventions
(ISBN 978-1-78242-091-0)

Woodhead Publishing Series in Food Science, Technology and Nutrition: Number 272

Advances in Food and Beverage Labelling

Information and Regulations

Edited by

Paul Berryman

ELSEVIER

AMSTERDAM • BOSTON • CAMBRIDGE • HEIDELBERG
LONDON • NEW YORK • OXFORD • PARIS • SAN DIEGO
SAN FRANCISCO • SINGAPORE • SYDNEY • TOKYO
Woodhead Publishing is an imprint of Elsevier

WP
WOODHEAD
PUBLISHING

Woodhead Publishing is an imprint of Elsevier
80 High Street, Sawston, Cambridge, CB22 3HJ, UK
225 Wyman Street, Waltham, MA 02451, USA
Langford Lane, Kidlington, OX5 1GB, UK

British Library Cataloguing-in-Publication Data
A catalogue record for this book is available from the British Library

Library of Congress Control Number: 2014944422

ISBN 978-1-78242-085-9 (print)
ISBN 978-1-78242-093-4 (online)

For information on all Woodhead Publishing publications
visit our website at http://store.elsevier.com/

Typeset by Newgen Knowledge Works Pvt Ltd, India
Printed and bound in the United Kingdom

Contents

List of contributors

N. Baldwin Intertek Scientific & Regulatory Consultancy, Farnborough, UK

P. Berryman Berryman Food Science Ltd, Cheriton, UK

M. Dean Queen's University Belfast, Belfast, UK

L. Fitzpatrick Ashbury Labelling, London, UK

B. Gallani Food and Drink Federation (FDF), London, UK

I.N. Hancock Trading Standards Service, Dorset County Council, Dorchester, UK

C. Hodgkins University of Surrey, Guildford, UK

L. Lähteenmäki MAPP Aarhus University, Aarhus, Denmark

S. Osborn Leatherhead Food Research, Leatherhead, UK

M.M. Raats University of Surrey, Guildford, UK

G.A. Skinner Insignia Technologies Ltd, Newhouse, UK

M. Spence Queen's University Belfast, Belfast, UK

J. Troth Assessment Compliance Expertise Ltd, Lichfield, UK

Woodhead Publishing Series in Food Science, Technology and Nutrition

1 **Chilled foods: A comprehensive guide**
Edited by C. Dennis and M. Stringer
2 **Yoghurt: Science and technology**
A. Y. Tamime and R. K. Robinson
3 **Food processing technology: Principles and practice**
P. J. Fellows
4 **Bender's dictionary of nutrition and food technology Sixth edition**
D. A. Bender
5 **Determination of veterinary residues in food**
Edited by N. T. Crosby
6 **Food contaminants: Sources and surveillance**
Edited by C. Creaser and R. Purchase
7 **Nitrates and nitrites in food and water**
Edited by M. J. Hill
8 **Pesticide chemistry and bioscience: The food-environment challenge**
Edited by G. T. Brooks and T. Roberts
9 **Pesticides: Developments, impacts and controls**
Edited by G. A. Best and A. D. Ruthven
10 **Dietary fibre: Chemical and biological aspects**
Edited by D. A. T. Southgate, K. W. Waldron, I. T. Johnson and G. R. Fenwick
11 **Vitamins and minerals in health and nutrition**
M. Tolonen
12 **Technology of biscuits, crackers and cookies Second edition**
D. Manley
13 **Instrumentation and sensors for the food industry**
Edited by E. Kress-Rogers
14 **Food and cancer prevention: Chemical and biological aspects**
Edited by K. W. Waldron, I. T. Johnson and G. R. Fenwick
15 **Food colloids: Proteins, lipids and polysaccharides**
Edited by E. Dickinson and B. Bergenstahl
16 **Food emulsions and foams**
Edited by E. Dickinson
17 **Maillard reactions in chemistry, food and health**
Edited by T. P. Labuza, V. Monnier, J. Baynes and J. O'Brien
18 **The Maillard reaction in foods and medicine**
Edited by J. O'Brien, H. E. Nursten, M. J. Crabbe and J. M. Ames

19 **Encapsulation and controlled release**
Edited by D. R. Karsa and R. A. Stephenson
20 **Flavours and fragrances**
Edited by A. D. Swift
21 **Feta and related cheeses**
Edited by A. Y. Tamime and R. K. Robinson
22 **Biochemistry of milk products**
Edited by A. T. Andrews and J. R. Varley
23 **Physical properties of foods and food processing systems**
M. J. Lewis
24 **Food irradiation: A reference guide**
V. M. Wilkinson and G. Gould
25 **Kent's technology of cereals: An introduction for students of food science and agriculture Fourth edition**
N. L. Kent and A. D. Evers
26 **Biosensors for food analysis**
Edited by A. O. Scott
27 **Separation processes in the food and biotechnology industries: Principles and applications**
Edited by A. S. Grandison and M. J. Lewis
28 **Handbook of indices of food quality and authenticity**
R. S. Singhal, P. K. Kulkarni and D. V. Rege
29 **Principles and practices for the safe processing of foods**
D. A. Shapton and N. F. Shapton
30 **Biscuit, cookie and cracker manufacturing manuals Volume 1: Ingredients**
D. Manley
31 **Biscuit, cookie and cracker manufacturing manuals Volume 2: Biscuit doughs**
D. Manley
32 **Biscuit, cookie and cracker manufacturing manuals Volume 3: Biscuit dough piece forming**
D. Manley
33 **Biscuit, cookie and cracker manufacturing manuals Volume 4: Baking and cooling of biscuits**
D. Manley
34 **Biscuit, cookie and cracker manufacturing manuals Volume 5: Secondary processing in biscuit manufacturing**
D. Manley
35 **Biscuit, cookie and cracker manufacturing manuals Volume 6: Biscuit packaging and storage**
D. Manley
36 **Practical dehydration Second edition**
M. Greensmith
37 **Lawrie's meat science Sixth edition**
R. A. Lawrie

38 **Yoghurt: Science and technology Second edition**
A. Y. Tamime and R. K. Robinson
39 **New ingredients in food processing: Biochemistry and agriculture**
G. Linden and D. Lorient
40 **Benders' dictionary of nutrition and food technology Seventh edition**
D. A. Bender and A. E. Bender
41 **Technology of biscuits, crackers and cookies Third edition**
D. Manley
42 **Food processing technology: Principles and practice Second edition**
P. J. Fellows
43 **Managing frozen foods**
Edited by C. J. Kennedy
44 **Handbook of hydrocolloids**
Edited by G. O. Phillips and P. A. Williams
45 **Food labelling**
Edited by J. R. Blanchfield
46 **Cereal biotechnology**
Edited by P. C. Morris and J. H. Bryce
47 **Food intolerance and the food industry**
Edited by T. Dean
48 **The stability and shelf-life of food**
Edited by D. Kilcast and P. Subramaniam
49 **Functional foods: Concept to product**
Edited by G. R. Gibson and C. M. Williams
50 **Chilled foods: A comprehensive guide Second edition**
Edited by M. Stringer and C. Dennis
51 **HACCP in the meat industry**
Edited by M. Brown
52 **Biscuit, cracker and cookie recipes for the food industry**
D. Manley
53 **Cereals processing technology**
Edited by G. Owens
54 **Baking problems solved**
S. P. Cauvain and L. S. Young
55 **Thermal technologies in food processing**
Edited by P. Richardson
56 **Frying: Improving quality**
Edited by J. B. Rossell
57 **Food chemical safety Volume 1: Contaminants**
Edited by D. Watson
58 **Making the most of HACCP: Learning from others' experience**
Edited by T. Mayes and S. Mortimore
59 **Food process modelling**
Edited by L. M. M. Tijskens, M. L. A. T. M. Hertog and B. M. Nicolaï

60 **EU food law: A practical guide**
Edited by K. Goodburn
61 **Extrusion cooking: Technologies and applications**
Edited by R. Guy
62 **Auditing in the food industry: From safety and quality to environmental and other audits**
Edited by M. Dillon and C. Griffith
63 **Handbook of herbs and spices Volume 1**
Edited by K. V. Peter
64 **Food product development: Maximising success**
M. Earle, R. Earle and A. Anderson
65 **Instrumentation and sensors for the food industry Second edition**
Edited by E. Kress-Rogers and C. J. B. Brimelow
66 **Food chemical safety Volume 2: Additives**
Edited by D. Watson
67 **Fruit and vegetable biotechnology**
Edited by V. Valpuesta
68 **Foodborne pathogens: Hazards, risk analysis and control**
Edited by C. de W. Blackburn and P. J. McClure
69 **Meat refrigeration**
S. J. James and C. James
70 **Lockhart and Wiseman's crop husbandry Eighth edition**
H. J. S. Finch, A. M. Samuel and G. P. F. Lane
71 **Safety and quality issues in fish processing**
Edited by H. A. Bremner
72 **Minimal processing technologies in the food industries**
Edited by T. Ohlsson and N. Bengtsson
73 **Fruit and vegetable processing: Improving quality**
Edited by W. Jongen
74 **The nutrition handbook for food processors**
Edited by C. J. K. Henry and C. Chapman
75 **Colour in food: Improving quality**
Edited by D. MacDougall
76 **Meat processing: Improving quality**
Edited by J. P. Kerry, J. F. Kerry and D. A. Ledward
77 **Microbiological risk assessment in food processing**
Edited by M. Brown and M. Stringer
78 **Performance functional foods**
Edited by D. Watson
79 **Functional dairy products Volume 1**
Edited by T. Mattila-Sandholm and M. Saarela
80 **Taints and off-flavours in foods**
Edited by B. Baigrie
81 **Yeasts in food**
Edited by T. Boekhout and V. Robert

82 **Phytochemical functional foods**
Edited by I. T. Johnson and G. Williamson
83 **Novel food packaging techniques**
Edited by R. Ahvenainen
84 **Detecting pathogens in food**
Edited by T. A. McMeekin
85 **Natural antimicrobials for the minimal processing of foods**
Edited by S. Roller
86 **Texture in food Volume 1: Semi-solid foods**
Edited by B. M. McKenna
87 **Dairy processing: Improving quality**
Edited by G. Smit
88 **Hygiene in food processing: Principles and practice**
Edited by H. L. M. Lelieveld, M. A. Mostert, B. White and J. Holah
89 **Rapid and on-line instrumentation for food quality assurance**
Edited by I. Tothill
90 **Sausage manufacture: Principles and practice**
E. Essien
91 **Environmentally-friendly food processing**
Edited by B. Mattsson and U. Sonesson
92 **Bread making: Improving quality**
Edited by S. P. Cauvain
93 **Food preservation techniques**
Edited by P. Zeuthen and L. Bøgh-Sørensen
94 **Food authenticity and traceability**
Edited by M. Lees
95 **Analytical methods for food additives**
R. Wood, L. Foster, A. Damant and P. Key
96 **Handbook of herbs and spices Volume 2**
Edited by K. V. Peter
97 **Texture in food Volume 2: Solid foods**
Edited by D. Kilcast
98 **Proteins in food processing**
Edited by R. Yada
99 **Detecting foreign bodies in food**
Edited by M. Edwards
100 **Understanding and measuring the shelf-life of food**
Edited by R. Steele
101 **Poultry meat processing and quality**
Edited by G. Mead
102 **Functional foods, ageing and degenerative disease**
Edited by C. Remacle and B. Reusens
103 **Mycotoxins in food: Detection and control**
Edited by N. Magan and M. Olsen

104 **Improving the thermal processing of foods**
Edited by P. Richardson
105 **Pesticide, veterinary and other residues in food**
Edited by D. Watson
106 **Starch in food: Structure, functions and applications**
Edited by A.-C. Eliasson
107 **Functional foods, cardiovascular disease and diabetes**
Edited by A. Arnoldi
108 **Brewing: Science and practice**
D. E. Briggs, P. A. Brookes, R. Stevens and C. A. Boulton
109 **Using cereal science and technology for the benefit of consumers: Proceedings of the 12th International ICC Cereal and Bread Congress, 24 – 26th May, 2004, Harrogate, UK**
Edited by S. P. Cauvain, L. S. Young and S. Salmon
110 **Improving the safety of fresh meat**
Edited by J. Sofos
111 **Understanding pathogen behaviour: Virulence, stress response and resistance**
Edited by M. Griffiths
112 **The microwave processing of foods**
Edited by H. Schubert and M. Regier
113 **Food safety control in the poultry industry**
Edited by G. Mead
114 **Improving the safety of fresh fruit and vegetables**
Edited by W. Jongen
115 **Food, diet and obesity**
Edited by D. Mela
116 **Handbook of hygiene control in the food industry**
Edited by H. L. M. Lelieveld, M. A. Mostert and J. Holah
117 **Detecting allergens in food**
Edited by S. Koppelman and S. Hefle
118 **Improving the fat content of foods**
Edited by C. Williams and J. Buttriss
119 **Improving traceability in food processing and distribution**
Edited by I. Smith and A. Furness
120 **Flavour in food**
Edited by A. Voilley and P. Etievant
121 **The Chorleywood bread process**
S. P. Cauvain and L. S. Young
122 **Food spoilage microorganisms**
Edited by C. de W. Blackburn
123 **Emerging foodborne pathogens**
Edited by Y. Motarjemi and M. Adams
124 **Benders' dictionary of nutrition and food technology Eighth edition**
D. A. Bender

125 **Optimising sweet taste in foods**
Edited by W. J. Spillane
126 **Brewing: New technologies**
Edited by C. Bamforth
127 **Handbook of herbs and spices Volume 3**
Edited by K. V. Peter
128 **Lawrie's meat science Seventh edition**
R. A. Lawrie in collaboration with D. A. Ledward
129 **Modifying lipids for use in food**
Edited by F. Gunstone
130 **Meat products handbook: Practical science and technology**
G. Feiner
131 **Food consumption and disease risk: Consumer–pathogen interactions**
Edited by M. Potter
132 **Acrylamide and other hazardous compounds in heat-treated foods**
Edited by K. Skog and J. Alexander
133 **Managing allergens in food**
Edited by C. Mills, H. Wichers and K. Hoffman-Sommergruber
134 **Microbiological analysis of red meat, poultry and eggs**
Edited by G. Mead
135 **Maximising the value of marine by-products**
Edited by F. Shahidi
136 **Chemical migration and food contact materials**
Edited by K. Barnes, R. Sinclair and D. Watson
137 **Understanding consumers of food products**
Edited by L. Frewer and H. van Trijp
138 **Reducing salt in foods: Practical strategies**
Edited by D. Kilcast and F. Angus
139 **Modelling microorganisms in food**
Edited by S. Brul, S. Van Gerwen and M. Zwietering
140 **Tamime and Robinson's Yoghurt: Science and technology Third edition**
A. Y. Tamime and R. K. Robinson
141 **Handbook of waste management and co-product recovery in food processing Volume 1**
Edited by K. W. Waldron
142 **Improving the flavour of cheese**
Edited by B. Weimer
143 **Novel food ingredients for weight control**
Edited by C. J. K. Henry
144 **Consumer-led food product development**
Edited by H. MacFie
145 **Functional dairy products Volume 2**
Edited by M. Saarela
146 **Modifying flavour in food**
Edited by A. J. Taylor and J. Hort

147 **Cheese problems solved**
Edited by P. L. H. McSweeney
148 **Handbook of organic food safety and quality**
Edited by J. Cooper, C. Leifert and U. Niggli
149 **Understanding and controlling the microstructure of complex foods**
Edited by D. J. McClements
150 **Novel enzyme technology for food applications**
Edited by R. Rastall
151 **Food preservation by pulsed electric fields: From research to application**
Edited by H. L. M. Lelieveld and S. W. H. de Haan
152 **Technology of functional cereal products**
Edited by B. R. Hamaker
153 **Case studies in food product development**
Edited by M. Earle and R. Earle
154 **Delivery and controlled release of bioactives in foods and nutraceuticals**
Edited by N. Garti
155 **Fruit and vegetable flavour: Recent advances and future prospects**
Edited by B. Brückner and S. G. Wyllie
156 **Food fortification and supplementation: Technological, safety and regulatory aspects**
Edited by P. Berry Ottaway
157 **Improving the health-promoting properties of fruit and vegetable products**
Edited by F. A. Tomás-Barberán and M. I. Gil
158 **Improving seafood products for the consumer**
Edited by T. Børresen
159 **In-pack processed foods: Improving quality**
Edited by P. Richardson
160 **Handbook of water and energy management in food processing**
Edited by J. Klemeš, R. Smith and J.-K. Kim
161 **Environmentally compatible food packaging**
Edited by E. Chiellini
162 **Improving farmed fish quality and safety**
Edited by Ø. Lie
163 **Carbohydrate-active enzymes**
Edited by K.-H. Park
164 **Chilled foods: A comprehensive guide Third edition**
Edited by M. Brown
165 **Food for the ageing population**
Edited by M. M. Raats, C. P. G. M. de Groot and W. A Van Staveren
166 **Improving the sensory and nutritional quality of fresh meat**
Edited by J. P. Kerry and D. A. Ledward
167 **Shellfish safety and quality**
Edited by S. E. Shumway and G. E. Rodrick
168 **Functional and speciality beverage technology**
Edited by P. Paquin

169 **Functional foods: Principles and technology**
M. Guo
170 **Endocrine-disrupting chemicals in food**
Edited by I. Shaw
171 **Meals in science and practice: Interdisciplinary research and business applications**
Edited by H. L. Meiselman
172 **Food constituents and oral health: Current status and future prospects**
Edited by M. Wilson
173 **Handbook of hydrocolloids Second edition**
Edited by G. O. Phillips and P. A. Williams
174 **Food processing technology: Principles and practice Third edition**
P. J. Fellows
175 **Science and technology of enrobed and filled chocolate, confectionery and bakery products**
Edited by G. Talbot
176 **Foodborne pathogens: Hazards, risk analysis and control Second edition**
Edited by C. de W. Blackburn and P. J. McClure
177 **Designing functional foods: Measuring and controlling food structure breakdown and absorption**
Edited by D. J. McClements and E. A. Decker
178 **New technologies in aquaculture: Improving production efficiency, quality and environmental management**
Edited by G. Burnell and G. Allan
179 **More baking problems solved**
S. P. Cauvain and L. S. Young
180 **Soft drink and fruit juice problems solved**
P. Ashurst and R. Hargitt
181 **Biofilms in the food and beverage industries**
Edited by P. M. Fratamico, B. A. Annous and N. W. Gunther
182 **Dairy-derived ingredients: Food and neutraceutical uses**
Edited by M. Corredig
183 **Handbook of waste management and co-product recovery in food processing Volume 2**
Edited by K. W. Waldron
184 **Innovations in food labelling**
Edited by J. Albert
185 **Delivering performance in food supply chains**
Edited by C. Mena and G. Stevens
186 **Chemical deterioration and physical instability of food and beverages**
Edited by L. H. Skibsted, J. Risbo and M. L. Andersen
187 **Managing wine quality Volume 1: Viticulture and wine quality**
Edited by A. G. Reynolds
188 **Improving the safety and quality of milk Volume 1: Milk production and processing**
Edited by M. Griffiths

189 **Improving the safety and quality of milk Volume 2: Improving quality in milk products**
Edited by M. Griffiths
190 **Cereal grains: Assessing and managing quality**
Edited by C. Wrigley and I. Batey
191 **Sensory analysis for food and beverage quality control: A practical guide**
Edited by D. Kilcast
192 **Managing wine quality Volume 2: Oenology and wine quality**
Edited by A. G. Reynolds
193 **Winemaking problems solved**
Edited by C. E. Butzke
194 **Environmental assessment and management in the food industry**
Edited by U. Sonesson, J. Berlin and F. Ziegler
195 **Consumer-driven innovation in food and personal care products**
Edited by S. R. Jaeger and H. MacFie
196 **Tracing pathogens in the food chain**
Edited by S. Brul, P. M. Fratamico and T. A. McMeekin
197 **Case studies in novel food processing technologies: Innovations in processing, packaging, and predictive modelling**
Edited by C. J. Doona, K. Kustin and F. E. Feeherry
198 **Freeze-drying of pharmaceutical and food products**
T.-C. Hua, B.-L. Liu and H. Zhang
199 **Oxidation in foods and beverages and antioxidant applications Volume 1: Understanding mechanisms of oxidation and antioxidant activity**
Edited by E. A. Decker, R. J. Elias and D. J. McClements
200 **Oxidation in foods and beverages and antioxidant applications Volume 2: Management in different industry sectors**
Edited by E. A. Decker, R. J. Elias and D. J. McClements
201 **Protective cultures, antimicrobial metabolites and bacteriophages for food and beverage biopreservation**
Edited by C. Lacroix
202 **Separation, extraction and concentration processes in the food, beverage and nutraceutical industries**
Edited by S. S. H. Rizvi
203 **Determining mycotoxins and mycotoxigenic fungi in food and feed**
Edited by S. De Saeger
204 **Developing children's food products**
Edited by D. Kilcast and F. Angus
205 **Functional foods: Concept to product Second edition**
Edited by M. Saarela
206 **Postharvest biology and technology of tropical and subtropical fruits Volume 1: Fundamental issues**
Edited by E. M. Yahia

207 **Postharvest biology and technology of tropical and subtropical fruits Volume 2: Açai to citrus**
Edited by E. M. Yahia
208 **Postharvest biology and technology of tropical and subtropical fruits Volume 3: Cocona to mango**
Edited by E. M. Yahia
209 **Postharvest biology and technology of tropical and subtropical fruits Volume 4: Mangosteen to white sapote**
Edited by E. M. Yahia
210 **Food and beverage stability and shelf life**
Edited by D. Kilcast and P. Subramaniam
211 **Processed Meats: Improving safety, nutrition and quality**
Edited by J. P. Kerry and J. F. Kerry
212 **Food chain integrity: A holistic approach to food traceability, safety, quality and authenticity**
Edited by J. Hoorfar, K. Jordan, F. Butler and R. Prugger
213 **Improving the safety and quality of eggs and egg products Volume 1**
Edited by Y. Nys, M. Bain and F. Van Immerseel
214 **Improving the safety and quality of eggs and egg products Volume 2**
Edited by F. Van Immerseel, Y. Nys and M. Bain
215 **Animal feed contamination: Effects on livestock and food safety**
Edited by J. Fink-Gremmels
216 **Hygienic design of food factories**
Edited by J. Holah and H. L. M. Lelieveld
217 **Manley's technology of biscuits, crackers and cookies Fourth edition**
Edited by D. Manley
218 **Nanotechnology in the food, beverage and nutraceutical industries**
Edited by Q. Huang
219 **Rice quality: A guide to rice properties and analysis**
K. R. Bhattacharya
220 **Advances in meat, poultry and seafood packaging**
Edited by J. P. Kerry
221 **Reducing saturated fats in foods**
Edited by G. Talbot
222 **Handbook of food proteins**
Edited by G. O. Phillips and P. A. Williams
223 **Lifetime nutritional influences on cognition, behaviour and psychiatric illness**
Edited by D. Benton
224 **Food machinery for the production of cereal foods, snack foods and confectionery**
L.-M. Cheng
225 **Alcoholic beverages: Sensory evaluation and consumer research**
Edited by J. Piggott

226 **Extrusion problems solved: Food, pet food and feed**
M. N. Riaz and G. J. Rokey
227 **Handbook of herbs and spices Second edition Volume 1**
Edited by K. V. Peter
228 **Handbook of herbs and spices Second edition Volume 2**
Edited by K. V. Peter
229 **Breadmaking: Improving quality Second edition**
Edited by S. P. Cauvain
230 **Emerging food packaging technologies: Principles and practice**
Edited by K. L. Yam and D. S. Lee
231 **Infectious disease in aquaculture: Prevention and control**
Edited by B. Austin
232 **Diet, immunity and inflammation**
Edited by P. C. Calder and P. Yaqoob
233 **Natural food additives, ingredients and flavourings**
Edited by D. Baines and R. Seal
234 **Microbial decontamination in the food industry: Novel methods and applications**
Edited by A. Demirci and M.O. Ngadi
235 **Chemical contaminants and residues in foods**
Edited by D. Schrenk
236 **Robotics and automation in the food industry: Current and future technologies**
Edited by D. G. Caldwell
237 **Fibre-rich and wholegrain foods: Improving quality**
Edited by J. A. Delcour and K. Poutanen
238 **Computer vision technology in the food and beverage industries**
Edited by D.-W. Sun
239 **Encapsulation technologies and delivery systems for food ingredients and nutraceuticals**
Edited by N. Garti and D. J. McClements
240 **Case studies in food safety and authenticity**
Edited by J. Hoorfar
241 **Heat treatment for insect control: Developments and applications**
D. Hammond
242 **Advances in aquaculture hatchery technology**
Edited by G. Allan and G. Burnell
243 **Open innovation in the food and beverage industry**
Edited by M. Garcia Martinez
244 **Trends in packaging of food, beverages and other fast-moving consumer goods (FMCG)**
Edited by N. Farmer
245 **New analytical approaches for verifying the origin of food**
Edited by P. Brereton

246 **Microbial production of food ingredients, enzymes and nutraceuticals**
Edited by B. McNeil, D. Archer, I. Giavasis and L. Harvey
247 **Persistent organic pollutants and toxic metals in foods**
Edited by M. Rose and A. Fernandes
248 **Cereal grains for the food and beverage industries**
E. Arendt and E. Zannini
249 **Viruses in food and water: Risks, surveillance and control**
Edited by N. Cook
250 **Improving the safety and quality of nuts**
Edited by L. J. Harris
251 **Metabolomics in food and nutrition**
Edited by B. C. Weimer and C. Slupsky
252 **Food enrichment with omega-3 fatty acids**
Edited by C. Jacobsen, N. S. Nielsen, A. F. Horn and A.-D. M. Sørensen
253 **Instrumental assessment of food sensory quality: A practical guide**
Edited by D. Kilcast
254 **Food microstructures: Microscopy, measurement and modelling**
Edited by V. J. Morris and K. Groves
255 **Handbook of food powders: Processes and properties**
Edited by B. R. Bhandari, N. Bansal, M. Zhang and P. Schuck
256 **Functional ingredients from algae for foods and nutraceuticals**
Edited by H. Domínguez
257 **Satiation, satiety and the control of food intake: Theory and practice**
Edited by J. E. Blundell and F. Bellisle
258 **Hygiene in food processing: Principles and practice Second edition**
Edited by H. L. M. Lelieveld, J. Holah and D. Napper
259 **Advances in microbial food safety Volume 1**
Edited by J. Sofos
260 **Global safety of fresh produce: A handbook of best practice, innovative commercial solutions and case studies**
Edited by J. Hoorfar
261 **Human milk biochemistry and infant formula manufacturing technology**
Edited by M. Guo
262 **High throughput screening for food safety assessment: Biosensor technologies, hyperspectral imaging and practical applications**
Edited by A. K. Bhunia, M. S. Kim and C. R. Taitt
263 **Foods, nutrients and food ingredients with authorised EU health claims: Volume 1**
Edited by M. J. Sadler
264 **Handbook of food allergen detection and control**
Edited by S. Flanagan
265 **Advances in fermented foods and beverages: Improving quality, technologies and health benefits**
Edited by W. Holzapfel

266 **Metabolomics as a tool in nutrition research**
Edited by J.-L. Sébédio and L. Brennan
267 **Dietary supplements: Safety, efficacy and quality**
Edited by K. Berginc and S. Kreft
268 **Grapevine breeding programs for the wine industry: Traditional and molecular technologies**
Edited by A. G. Reynolds
269 **Handbook of natural antimicrobials for food safety and quality**
Edited by M. Taylor
270 **Managing and preventing obesity: Behavioural factors and dietary interventions**
Edited by T. P. Gill
271 **Electron beam pasteurization and complementary food processing technologies**
Edited by S. D. Pillai and S. Shayanfar
272 **Advances in food and beverage labelling: Information and regulations**
Edited by P. Berryman
273 **Flavour development, analysis and perception in food and beverages**
Edited by J. K. Parker, S. Elmore and L. Methven
274 **Rapid sensory profiling techniques and related methods: Applications in new product development and consumer research**
Edited by J. Delarue, J. B. Lawlor and M. Rogeaux
275 **Advances in microbial food safety: Volume 2**
Edited by J. Sofos
276 **Handbook of antioxidants in food preservation**
Edited by F. Shahidi
277 **Lockhart and Wiseman's crop husbandry including grassland: Ninth edition**
H. J. S. Finch, A. M. Samuel and G. P. F. Lane
278 **Global legislation for food contact materials: Processing, storage and packaging**
Edited by J. S. Baughan
279 **Colour additives for food and beverages: Development, safety and applications**
Edited by M. Scotter
280 **A complete course in canning and related processes 14th Edition: Volume 1**
Revised by S. Featherstone
281 **A complete course in canning and related processes 14th Edition: Volume 2**
Revised by S. Featherstone
282 **A complete course in canning and related processes 14th Edition: Volume 3**
Revised by S. Featherstone

Preface

As a Consultant in Food Science and Food Law, I am delighted to edit this book on advances in food and beverage labelling. Over the last 30 years, I have worked as a Public Analyst, Head of Trading Standards and Director of a food industry research association, so I have acted as both gamekeeper and poacher when it comes to food labelling! I therefore understand the challenges that both industry and enforcement bodies face when interpreting complex food labelling laws.

In 2012, when I was first approached by Woodhead Publishing to edit a book on food labelling, I was CEO of Leatherhead Food Research. With a large Regulatory Advice team serving the global food industry, it was our business to know all about food labelling, so I readily accepted the opportunity. We were already planning for the impending Food Information for Consumers Regulations, which was one of the most important pieces of EU Food Labelling legislation for over 20 years. EU Regulation No. 1169/2011 overhauls and streamlines all food information for consumers, including quantitative ingredient declarations (QUID), allergen labelling, durability indications, country of origin and nutrition information. I predicted that, between 2014 and 2016, virtually all EU food labels would need to be updated. Two years later, I think my predictions are correct.

So it is timely for this book to be published at the same time as the implementation of these far-reaching new regulations. I describe them in more detail in Chapter 1.

This book is intended for anyone interested in food labelling, nutrition information and innovations relating to environmental, genetically modified (GM), ethical and even smart labelling (e.g. labels that change colour when food deteriorates). It will be of particular interest to food scientists and regulatory professionals working for food manufacturers and retailers, so that they are aware of new legislative trends including comparisons between EU and USA food laws.

Enforcement officers in Central and Local Government will be just as interested. Trading Standards Officers and Environmental Health professionals will particularly enjoy Chapters 3 and 4 covering enforcement bodies and the latest enforcement policies and codes in the UK.

As a visiting Professor at the University of Reading, I know that this book will also appeal to academic audiences, particularly those teaching courses on food science, food technology, nutrition and food law and, of course, the students of BSc, MSc and PhD programmes.

The book is an improvement on the current literature because for the first time we have experts from industry, trade bodies, government and academia all giving the very latest insights into food labelling for both EU and USA markets.

There are three main parts to the book.

Part One covers general trends in EU and USA food and beverage labelling regulations, including two excellent chapters on enforcement bodies, policies and codes of practice.

Part Two focuses on nutrition and health claims, which have become increasingly important over recent years. There are useful comparisons between EU and USA laws, the latest information on "front-of-pack" labelling, and an interesting chapter on consumer understanding of nutrition labelling. This is especially relevant because health claims and nutrition information must be easy for consumers to use and enable fair comparisons of similar products.

Part Three covers newer developments, such as GM and ethical and environmental labelling, often not compulsory but very relevant to certain consumers who are just as interested in these factors as nutrition and price. Chapters on smart labelling and clean labels – those with only natural ingredients for example – round off the book.

Readers can dip into the book for reference purposes, read any chapter as a standalone treatise, or read it from cover to cover if food labelling is an integral part of their day-to-day job.

I must finish by thanking all of the contributors to the book, each a recognised expert in their field. I also thank Woodhead Publishing for all the guidance on pulling together such an eclectic book. Finally, thanks to my wife Margaret and sons Andy and Alex (budding nutritionist) for their encouragement – when I could have been engaged in family activities rather than editing this book!

Paul Berryman

Part One

General trends in regulation and enforcement of food labelling

The EU food information for consumers regulation

1

P. Berryman
Berryman Food Science Ltd, Cheriton, UK

1.1 Introduction

In 2008, the European Commission's Directorate-General for Consumer and Health issued a proposal to combine two main Directives (Directive 2000/13/EC and Directive 90/496/EEC) into one Framework Regulation 1169/2011 (Official Journal of the European Union, 2011). This followed an evaluation of all EU legislation on food labelling by the European Commission.

The new regulation establishes a legal framework in the EU with regard to information related to foodstuffs provided to consumers by food business operators (FBO) at all stages of the food chain. Its aim is to *"serve the interests of the internal market by simplifying the law, ensuring legal certainty and reducing administrative burden, and benefit citizens by requiring clear, comprehensible and legible labelling of foods."* The regulation includes general principles and responsibilities, as well as requirements governing food information. The provisions apply to all foods intended for the final consumer, including foods delivered by mass caterers, and foods intended for supply to mass caterers.

In the UK, EU Regulation 1169/2011 is known as the Food Information Regulation 2013 (Draft Statutory Instrument, 2013) and will replace the Food Labelling Regulations 1996.

The new regulation will help to make food labels clearer and also improve nutritional and allergen information to consumers. Although many provisions will remain unchanged, there are significant changes that FBOs must address.

The changes include:

- mandatory nutrition labelling on pre-packaged food;
- more country of origin labelling;
- improved date marking (including date of first freezing);
- clarity and legibility of food information (including minimum font size);
- labelling of non-pre-packed foods;
- allergen information, including on food sold loose and in restaurants and catering outlets.

It also introduces improvement notices which can be issued by trading standards officers for failure to comply with some sections of the regulations. Failure to comply with the improvement notice is a criminal offence.

Advances in Food and Beverage Labelling. http://dx.doi.org/10.1533/9781782420934.1.3

The regulation entered into force on 13 December 2011 and states that the provisions do not have to be applied until the following dates:

- 1 January 2014 – For the rules on the composition and labelling of minced meat.
- 13 December 2014 – For the general labelling rules (including allergens), and if provided, a nutrition declaration must use the format set out in regulations.
- 13 December 2016 – For the rules on mandatory nutrition declaration needed for most pre-packed food.

FBOs can choose to comply earlier if they wish, for example when redesigning labels.

In my opinion, the best guidance notes on the implementation of the regulations were produced by Food Drink Europe, the European Trade Body (Food Drink Europe, 2013). As well as giving worked examples, it also incorporates the interpretation given in the EU Questions and Answers Document on Regulation (EU) 1169/2011 available on the EC website. This gives very useful guidance not mentioned in the regulations. For example it clarifies that salt declarations must be calculated from the total sodium × 2.5, even if some of the sodium comes from ingredients other than sodium chloride (e.g. sodium ascorbate or sodium bicarbonate).

The UK Government's Department for Food & Rural Affairs (Defra) has issued a guide to compliance which provides further useful information (Defra, 2012).

1.2 General requirements and responsibilities

Chapter III of the Regulations give an overview of the general principles and requirements governing the provision of food information, including fair information practices and responsibilities of FBOs. The details are given in the regulations but in summary:

The basic requirement is that any food intended for supply to the final consumer or to mass caterers shall be accompanied by food information that is *fair*:

Information must not be misleading, particularly:

- as to the characteristics of the food and, in particular, as to its nature, identity, properties, composition, quantity, durability, country of origin or place of provenance, method of manufacture or production;
- by attributing to the food effects or properties which it does not possess;
- by suggesting that the food possesses special characteristics when in fact all similar foods possess such characteristics, in particular by specifically emphasising the presence or absence of certain ingredients and/or nutrients;
- by suggesting, by means of the appearance, the description or pictorial representations, the presence of a particular food or an ingredient, while in reality a component naturally present or an ingredient normally used in that food has been substituted with a different component or a different ingredient.

Food information shall be accurate, clear and easy to understand for the consumer.

Food information shall not attribute to any food the property of preventing, treating or curing a human disease, nor refer to such properties. (Note: subject to derogations for natural mineral waters and foods for particular nutritional uses.)

Food information includes that given via advertising, presentation (shape, appearance, packaging) and settings, as well as labels.

The regulations make it clear that the FBO is responsible for the presence and accuracy of food information (and any changes they make to food information). The FBO is the operator under whose name, or business name, the food is marketed or, if that operator is not established in the Union, is the importer into the Union market. This also means that restaurant staff need to know about relevant food information, for example if meals contain any allergens.

1.3 Mandatory food information

This section gives a detailed consideration of the mandatory labelling requirements.

1.3.1 Mandatory particulars

FBOs *must* indicate the following particulars:

- the name of the food;
- the list of ingredients;
- any ingredient or processing aid (listed in Annex II) causing allergies or intolerances;
- the quantity of certain ingredients or categories of ingredients;
- the net quantity of the food;
- the date of minimum durability or the "use by" date;
- any special storage conditions and/or conditions of use;
- the name or business name and address of the FBO (Article 8(1));
- the country of origin or place of provenance where provided for in Article 26;
- instructions for use where it would be difficult to make appropriate use of the food in the absence of such instructions;
- with respect to beverages containing more than 1.2% by volume of alcohol, the actual alcoholic strength by volume;
- a nutrition declaration.

The above particulars will usually be indicated with words and numbers, but may additionally be expressed by means of pictograms or symbols.

1.3.2 Additional particulars for specific foods

This refers to Annex III, which lists foods for which the labelling must include one or more additional particulars. Examples include packaging gases, sweeteners, confectionery containing glycyrrhizinic acid (licorice), caffeine, phytosterols and frozen meat or fish.

1.3.3 Presentation of mandatory particulars

Mandatory food information must be available and easily accessible for all foods. For pre-packed foods it must be on the package or an attached label. The regulations also indicate that other means of provision of information may be adopted if useful for consumer understanding. In the case of non-pre-packed food, except for allergen labelling, Member States can adopt different measures in certain circumstances (see Article 44).

One of the most controversial changes to the regulations relates to the requirement for a minimum font size for mandatory particulars. Although this is of obvious benefit to consumers who have difficulty reading very small print, it is troublesome for some FBOs. They argue that this requirement leaves very little room on the packaging for advertising information and trade names, etc. Nevertheless, the requirements are quite clear:

- Mandatory food information shall be marked in a conspicuous place in such a way as to be easily visible, clearly legible and, where appropriate, indelible. It shall not in any way be hidden, obscured, detracted from or interrupted by any other written or pictorial matter or any other intervening material.
- The mandatory particulars shall be printed on the package or on the label in such a way as to ensure clear legibility, in characters using a font size where the x-height, as defined in Annex IV, is equal to or greater than 1.2 mm.
- In case of packaging or containers, the largest surface of which has an area of less than 80 cm^2, the x-height of the font size referred to in Paragraph 2 shall be equal to or greater than 0.9 mm.
- Certain particulars (name of food, net quantity and alcohol by volume) shall appear in the same field of vision (except for tiny packages and some reusable glass bottles, e.g. milk).

1.3.4 Name of the food

If the food has a name prescribed by law, then this must be used, e.g. coffee, chocolate, sugar or jam. If not, then the customary name should be used. This is the name normally accepted by customers (in a given EU Member State) without needing any further explanation, e.g. in UK, Bakewell tart or fish fingers. For all other foods, a descriptive name should be used, e.g. cereal bar.

Annex VI of the Regulations specifies a whole range of extra information needed to accompany the name of the food. This includes its physical condition (powdered, refrozen, freeze dried, smoked, concentrated, etc.), whether it has been irradiated (must state "irradiated" or "treated with ionising radiation") and, for a fish or meat product, whether it has added water or has been chopped and reshaped or reformed.

There are also specific requirements for minced meat, depending on the amount of fat and collagen (connective tissue) it contains. For example, *lean minced meat* cannot contain more than 7% fat nor have a collagen/meat protein ratio higher than 12%.

1.3.5 Indication and designation of ingredients

Foods should be labelled with an ingredients list, stating each ingredient in descending order by weight.

Articles 18, 19 and 20 describe the format of an ingredients list: exemptions (e.g. fresh fruit and vegetables, carbonated water, fermented vinegar, cheese and butter), and which food constituents do not need to be given in the list (e.g. processing aids, carriers and water used for rehydration of a dried ingredient).

Any engineered nanomaterials must have "nano" in brackets after its name in the list. This nano requirement is quite controversial, because many foods contain natural nanoparticles e.g. salt and nanoemulsions. It is feared that "nano" labelling will cause unnecessary consumer concern and product rejection. Some beverage and dairy products marketed in China and Singapore proudly claim the presence of nano-calcium as a health benefit, based on the premise that very small particles of calcium are absorbed into the body more readily than larger ones, e.g. Da Ren Beverage, Anlene Gold (Fonterra).

Annex VII gives further requirements:

- Part A: clarifies the indication of ingredients in descending order by weight. Examples include vegetable oils, fruit, vegetables, mushrooms, herbs and spices that can be declared "in varying, (or variable), proportions" to allow for seasonality.
- Part B: allows categories of ingredients to be used rather than the specific name for some ingredients, including refined animal oils and fats, flour mixtures, fish, cheese, herbs, gum base, sugar, cocoa butter, wine and meat.
- Part C: this gives a list of 24 categories of food additive that can appear on an ingredients list, followed by the specific name or "E number," e.g. acid, colour, preservative.
- Part D: covers flavourings; smoke flavourings; and use of the word "natural" to describe flavourings and the additives quinine and caffeine as flavourings.
- Part E: describes how compound (multi-component) ingredients should be listed.

1.3.6 Quantitative ingredients declaration (QUID)

Article 22 and Annex VIII relate to the requirement to give a quantitative declaration of ingredients (or categories) where:

- They appear in the food or are usually associated by the consumer with that name, e.g. chocolate biscuit needs to indicate % chocolate.
- They are emphasised by words, pictures or graphics on the label, e.g. a picture of a strawberry on a dessert would require % strawberry to be indicated.
- They are essential to characterise a food and to distinguish it from other similar products.

Two interesting points are: (a) quantities of volatile ingredient (e.g. brandy) should be expressed as a % of the finished product – even though much of it may have evaporated during cooking, and (b) for dehydrated products such as soup powder, the ingredient proportions can be expressed on the basis of their proportion by weight of the reconstituted product after addition of water.

The requirements have remained the same as those given in Directive No. 2000/13/EC (Official Journal of the European Union, 2000). EC guidance III/5260-rev5/98 still applies (EC Europa, 1998).

1.3.7 Allergen declaration

Annex II lists allergens that must be highlighted in the ingredients list (in bold, or in different colour or underlined, for example). This is a big change for food manufacturers, who previously tended to list allergens in an allergen box. Under these regulations, allergen information may not be repeated elsewhere on the labelling, except for signposting the allergen information in the ingredients list. This requirement has had mixed reviews from commentators. On the one hand, it standardises all allergen information, so a sufferer knows to look at the ingredient list for allergen guidance. On the other hand, the information is less obvious than when it was front-of-pack, so may be overlooked.

If an allergen is used in the preparation of a food that does not need an ingredients list, then the food must bear the declaration "contains" followed by the name of the allergen. So, although wine does not need an ingredients list, it may need the declaration "contains sulphites." These regulations only cover deliberately added ingredients so do not address allergen cross-contamination or "free from" claims.

Annex II lists 14 substances that cause allergies or intolerances, as follows:

- Cereals containing gluten (wheat, rye, barley, oats, spelt, Kamut)
- Crustaceans (prawn, shrimp, etc.)
- Eggs
- Fish and products except fish gelatine
- Peanuts
- Soybeans
- Milk
- Nuts (almond, hazelnut, walnut, cashew, pecan, brazil, pistachio, macadamia)
- Celery
- Mustard
- Sesame
- Sulphur dioxide and sulphites at concentrations over 10 mg/kg or 10 mg/L
- Lupin
- Molluscs

In the UK, The Food Standards Agency has issued some advise with regards to allergen labelling (Food Standards Agency, 2013).

1.3.8 Durability indication

Article 24 and Annex X require most foods to be marked with a "best before" or "use by" date. The annex lists exempted foods, including fresh fruit and vegetables, wines, beverages containing at least 10% alcohol by volume, vinegar, cooking salt, solid sugar, chewing gums and certain pastries.

The "use by" date is for foods that are highly perishable and therefore likely to pose an immediate danger to health after a short period (e.g. ready to eat fish and meat products such as ham). Food on sale after this date is automatically deemed unsafe, so can lead to automatic prosecution. Some UK FBOs have complained because there is no longer a requirement for an enforcement officer to prove that the food is unsafe in order to prosecute under the General Food Regulations 2004.

The "best before" date is just an indication of the date before which the product is of best quality. Sale after this date is not an offence provided the product is safe to eat, e.g. stale biscuits.

The date of first freezing for meat preparations and unprocessed fishery products is also required. This is a new requirement that may surprise consumers, since some meat is frozen many months before sale.

1.3.9 Special storage conditions and usage instructions

Article 25 requires storage conditions and conditions of use to be given to ensure proper storage and use of the food. Examples include "once opened keep refrigerated and consume within 3 days" and "once defrosted do not refreeze."

1.3.10 Business name and address

Articles 8 and 9.1 (h) require an indication of the name or business name of the FBO. The name is the business responsible for the information on the labelling. It must be an FBO in the EU or, where the food is imported into the EU, the name of the importer. This is for traceability purposes so that consumers or enforcement officers can contact the business for any reason, including complaints.

1.3.11 Origin labelling

Article 26 and Annex XI require the origin or place of provenance of a food to be given:

- Where there is a mandatory vertical EU legislation (e.g. honey, olive oil, beef);
- For swine meat, sheep meat, goat meat as per Annex VI;
- Where failure to indicate its origin would mislead consumers, particularly if information accompanying the food might imply a certain origin, e.g. use of national flags or costumes.

The regulations also indicate that a range of impact assessments will be carried out in the future that may extend the mandatory origin labelling requirements.

1.3.12 Nutrition information

Nutrition declarations are covered by Section 3 of the Regulations (Articles 29–35). The requirements for nutrition labelling are extensive, and are only summarised here. For further details, I recommend the Defra Guidance notes (Defra, 2012) and Chapter 1 of Food Drink Europe Guidance on the Provision of Food Information to Consumers (Food Drink Europe, 2013).

Chapters 6 (Baldwin) and 7 (Dean) of this book also give details on health claims and front-of-pack nutrition labelling, respectively.

The nutrition labelling rules do not apply to food supplements or natural mineral waters, which have their own rules. Foods for particular nutritional foods (PARNUTS) also have vertical rules that have to be considered separately, e.g. infant formulae and special foods eaten for medical purposes.

Annex V lists 19 types of food that are exempted, including unprocessed single ingredient foods, herbs, spices, table top sweeteners, flavourings, chewing gums and food in very small packages.

The *mandatory nutrition declaration* must include:

- Energy value;
- Fat, saturates, carbohydrate, sugars, protein and salt.

There are also rules about other nutrient claims having to be in the same field of vision as the mandatory nutrition declaration, e.g. omega-3 fatty acids or beta glucans – units of measurement must also be appropriate.

FBOs may also add *voluntary nutrition information*, but if they do various rules apply. The voluntary nutrients are:

- Monounsaturates
- Polyunsaturates
- Polyols
- Starch
- Fibre
- Vitamins listed at Annex XII (i.e. A, D, E, K, C, B6, B12, thiamine, riboflavin, niacin, folic acid, biotin and pantothenic acid)
- Minerals listed at Annex XII (i.e. potassium, chloride, calcium, phosphorus, magnesium, iron, zinc, copper, manganese, fluoride, selenium, chromium, molybdenum and iodine).

There are also definitions of what constitutes a "significant amount" of vitamins and nutrients, reference intake data and energy conversion factors for different nutrients, e.g. protein and carbohydrates contain 4 kcal/g, fat contains 9, fibre contains 2 and alcohol contains 7.

It is interesting to note that fibre declarations are voluntary rather than mandatory, since many consumers are interested in dietary fibre for gut health purposes.

For energy value and nutrients, units of measurement are set out in Annexes XIII and XV. They should be expressed per 100 g or per 100 mL. Energy values must be given in both kilojoules and kilocalories. If vitamin or mineral declarations are made, then they must additionally be expressed as a percentage of the reference intakes set out in Annex XIII B. Nutrition information can also be given per portion.

The regulations also require a specific format for the presentation of nutrition information. Nutrients must be presented in a set order, either as a properly aligned table or, if space does not permit, in a linear format. The format rules are so that consumers can easily compare the nutrition on one food or brand with another.

1.3.13 Net quantity

Article 23 mandates net quantity declaration (with some exceptions and variations listed in Annex IX). Net quantity should be provided in units of volume in the case of liquid products and in units of mass in the case of other products. Member States have the ability to rule on how net quantity should be labelled i.e. by weight or volume. Interestingly, ice cream is declared by weight in some EU countries, notably France and Spain, and volume in others. In my opinion, it makes sense to declare net weight rather than volume, because ice cream contains incorporated air (the overrun), so volume comparisons can be very confusing for consumers. For example, an 850 g tub of ice cream would have a volume of 1.5 L at 50% overrun but a volume of 2 L at 100% overrun. The 2 L tub may look better value for money, but the customer is just buying extra air.

Annex IX also covers drained weight for foods in a liquid medium, such as vegetables in brine or fruits in syrup.

In the UK, these regulations are enforced under Business, Innovation and Skills (BIS) weights and measures regulations.

1.3.14 Instruction for use

Article 27 requires that instructions for use must be given so as to enable proper use to be made of the food. If a food requires cooking, then instructions on how to bake, roast, grill, fry, boil or microwave the food must be given. Manufacturers utilise independent laboratories to determine cooking instructions using calibrated cooking devices and special heat probes to ensure that microorganisms are destroyed during the cooking process.

1.3.15 Alcoholic strength

Article 28 and Annex XII require the declaration of alcoholic strength on beverages containing more than 1.2% by volume of alcohol. The figure should be followed by "%vol" and must not be more than one decimal place.

1.4 Distance selling

Article 14 requires that consumers receive the same information when buying food at a distance as they would when buying in a shop. All mandatory food information must be available before the purchase is completed and at the moment of delivery. It must be available at no extra cost, so for example, if the information is via a call centre, it must be via a Freephone line rather than a premium line. These requirements are particularly relevant due to the prevalence of online shopping.

1.5 Future trends

There is no doubt in my mind that the food information for consumers regulation will have far-reaching effects across the European Union. It will require virtually all food labels to be modified in some way. One of the key features of Regulation (EU) No. 1169/2011 is that almost all pre-packed foods will require mandatory nutrition labelling from 13 December 2016, in a new prescribed format. Even companies currently giving voluntary nutrition information will need to change their labels to comply with the new format and they have an even tighter deadline of 13 December 2014.

Although this will ultimately provide more consistency for the consumer, it presents a mammoth task for industry. I recommend that companies book their printers as soon as possible, to avoid the 2016 rush.

The regulation also allows for front-of-pack nutritional labelling using Guideline Daily Amounts (GDA) or traffic lights, etc. Since the regulations specify a minimum font size for mandatory information, there is arguably less space available for voluntary information, pictures and logos. Eventually, food labelling may require use of electronic systems, so that voluntary information can be viewed on mobile phones while shopping, via QR codes and similar devices. The increased trend for online shopping also presents opportunities for FBOs to present voluntary information to the consumer via websites and special portals.

Another key area that will be changed under these regulations is the labelling of allergens. The "allergen information box" will be phased out in deference to highlighting allergenic ingredients in the ingredients list. To comply with the letter of the law, to indicate the presence of gluten, the actual name of the cereal (wheat, barley etc.) needs to be highlighted in the list. This may confuse consumers familiar with the allergen box, so I foresee a steep learning curve ahead.

In the wake of the 2013 horsemeat scandal, I predict a greater emphasis on traceability of foods, provenance and country of origin. New techniques to trace foods and ingredients across the whole food chain using electronic tags, DNA markers and precision agriculture (linking crops and livestock to satellite and GPS tracking systems) will become more commonplace.

Finally, it is worth mentioning that consumer behaviour will continue to evolve, with internet access, smartphone ownership and social media all playing a part. Consumers will demand information when and where they want it – taking photos of food to identify brands, reviews, price comparisons and availability – and of course information beyond the scope of the humble food label.

References

Defra (2012), The Food Information Regulations 2013, Guide to Compliance, accessed on 26.5.14 at: https://www.gov.uk/government/uploads/system/uploads/attachment_data/file/82663/consult-fic-guidance-20121116.pdf.

Draft Statutory Instrument (2013), The Food Information Regulations 2013, accessed on 26.5.14 at: https://www.gov.uk/government/uploads/system/uploads/attachment_data/file/82668/consult-fic-draft-si-20121107.pdf.

EC Europa (1998), General guidelines for implementing the principle of Quantitative Ingredients Declaration (QUID) – Article 7 of Directive 79/112/EEC as amended by Directive 97/4/EC, (III/5260-rev5/98), accessed on 26.5.14 at: http://ec.europa.eu/food/food/labellingnutrition/resources/fl02_en.pdf.

Food Drink Europe (2013), Guidance on the Provision of Food Information to Consumers, Regulation (EU) No. 1169/2011, accessed on 26.5.14 at: http://www.fooddrinkeurope.eu/S=0/publication/guidance-on-the-provision-of-food-information-to-consumers/.

Food Standards Agency (2013), Advice on food allergen labelling, accessed on 26.5.14 at: http://multimedia.food.gov.uk/multimedia/pdfs/publication/allergy-leaflet.pdf.

Official Journal of the European Union (2000), Directive No. 2000/13/EC on the approximation of the laws of the Member States relating to the labelling, presentation and advertising of foodstuffs, accessed on 26.5.14 at: http://eur-lex.europa.eu/LexUriServ/LexUriServ.do?uri=OJ:L:2000:109:0029:0042:EN:PDF.

Official Journal of the European Union (2011), Regulation EU No 1169/2011 on the provision of food information to consumers, accessed on 26.5.14 at: http://eur-lex.europa.eu/LexUriServ/LexUriServ.do?uri=OJ:L:2011:304:0018:0063:EN:PDF.

Current regulation of food and beverage labelling in the USA

2

L. Fitzpatrick
Ashbury Labelling, London, UK

2.1 Introduction

In the United States, regulations surrounding provision of food information have evolved in a unique way. Although the stated principles of protecting public health and consumer interests are the same there are as they are in the European Union, there are many contrasts. Although both EU and US food information regulations are in an era of change, they are not necessarily growing closer together.

A review of the requirements of US food information regulation could easily constitute a book on its own, therefore this chapter can only provide an overview and will focus mainly on regulations developed and enforced by the Food and Drug Administration (FDA). This chapter looks at general requirements for food labelling. Since it is such a significant aspect of food labelling, there is a separate chapter (Chapter 5) that looks in more detail at labelling related to nutrition and health issues.

2.2 Regulatory oversight of labelling between government bodies

One complexity is the split in regulatory authority, sometimes depending upon the type of food, sometimes upon the subject area. We will begin with a brief introduction of the main players.

2.2.1 *FDA*

Among the Food FDA's stated areas of responsibility is assuring that foods sold in the United States are safe, wholesome and properly labelled, whether domestically produced, or imported (FDA, "FDA Basics for Industry"). FDA's remit covers approximately 80% of the food supply, as they have authority over all foods except those controlled by the United States Department of Agriculture and the Tax and Trade Bureau. FDA's regulatory scope also includes bottled waters, and dietary

Advances in Food and Beverage Labelling. http://dx.doi.org/10.1533/9781782420934.1.15

supplements, and their non-food responsibility includes drugs, tobacco products, cosmetics, animal feed and medicines, radiation-emitting devices, and vaccines.

For the purposes of this chapter, we will mainly be concerned with FDA's Office of Nutrition, Labeling, and Dietary Supplements. It is important to note that FDA does not pre-approve food labels placed on the US market, nor do its employees generally review labels upon request. FDA monitors compliance through random post-marketing surveillance, although priorities for certain product types or for claims are established. Label reviews generally arise in the event of a complaint by consumers or by competitors, or following an FDA on-site inspection.

Regulations promulgated by FDA are codified in Title 21 of the Code of Federal Regulations (21 CFR), which will be used as a reference in this chapter for simplicity.

2.2.2 United States Department of Agriculture (USDA)

The second agency with responsibility for food labelling is the USDA's Food Safety and Inspection Service (FSIS). USDA has authority over all meat and poultry products and processed egg products. FSIS develops and provides labelling guidance, policies, and inspection methods, and administers programs to protect consumers from misbranded and economically adulterated meat, poultry, and egg products which ensure that all labels are truthful and not misleading.

The regulatory split between FDA and USDA is not always obvious, as in general, mixed food products with more than 2% cooked meat or poultry (or 3% raw) are regulated by USDA; products with 2% or less cooked meat or poultry (3% or less raw) are regulated by FDA. For example, close-faced sandwiches are listed as FDA jurisdiction, while open-faced sandwiches are USDA. The meats regulated by USDA are those of cattle, sheep, swine, goats, horses, mules and other equines, ratites and squab, and any domesticated bird. By contrast, any non-specified red meats (bison, rabbits, game, zoo animals, all deer species, and wild birds are under FDA's jurisdiction (FDA, 2013).

This determination is important, as unlike FDA, USDA requires pre-approval of labels of products, and the Labeling Program and Delivery Division (LPDD) of FSIS evaluates approximately 60,000 label approval applications each year. USDA also administers the National Organic Program. USDA's regulations may be found in Title 9 of the Code of Federal Regulations (9 CFR).

2.2.3 TTB – Tax and Trade Bureau (Alcoholic Beverages)

Although alcoholic beverages (specifically spirits, wine, and malt beverages) are within FDA's import jurisdiction, their labelling and taxation are regulated by the Alcohol and Tobacco TTB rather than by FDA. TTB's labelling requirements for these products are laid down in separate regulations, which are codified in Title 27 of the Code of Federal Regulations (27 CFR). However, other beverages that contain less than 7% alcohol (for example, "wine coolers," hard ciders, and beers not made from malt) are instead subject to FDA labelling requirements as for most food products. TTB also requires prior approval of labels (as well as formulation) for products in its jurisdiction, and a detailed "Certificate of Label Approval" (COLA) submission procedure.

2.2.4 FTC – Federal Trade Commission

The Federal Trade Commission is the nation's consumer protection agency. The Federal Trade Commission Act ("FTC Act") empowers FTC to prohibit the false advertisement of "foods, drugs, and cosmetics." Although the definition of "advertisement" does not specifically include labelling, FTC has additional authority pursuant to Section 5 of the FTC Act to prevent "unfair or deceptive acts or practices in or affecting commerce." Therefore, FTC may take action against all unfair business practices, including false and misleading labelling of food products. The FTC Act makes "the dissemination of any false advertisement" an "unfair or deceptive practice for the purpose of inducing, or that is likely to induce, the purchase of food or having an effect on interstate commerce" (FTC, 1914). Therefore, this encompasses the concept of "food information," including that which appears on the label if it can be considered a form of advertising. FTC also lays down guidelines for the substantiation of claims. FTC also has statutory authority to prosecute offenders and, in some instances, damages. FTC may also require "corrective" advertising if necessary. FTC operates on Memoranda of Understanding in conjunction with FDA and other agencies to avoid duplication of their enforcement activities. FTC's regulations are found in Title 16 of the Code of Federal Regulations (16 CFR).

2.2.5 CBP – Customs and Border Protection

The U.S. Customs and Border Protection Agency, under the Department of Homeland Security, is in charge of matters of national security, administering entry requirements for imports under the Bioterrorism Act alongside FDA, but is also responsible for international trade matters. In particular, Country of Origin labelling requirements are established by the Tariff Act of 1930 and Customs Regulations, and also laid down in 19 CFR.

2.2.6 NIST (National Institute of Standards and Technology)

Other subject areas not controlled directly by FDA are date marking and metrology. Both of these are programs of the NIST. Part of the U.S. Department of Commerce, NIST is a physical science laboratory but its Office of Weights and Measures promotes uniformity in US weights and measures laws, regulations, and standards, and includes methods for checking of declared weights, which are not included in the CFR. It also publishes the Uniform Open Dating Regulation, discussed in Section 2.5 below.

2.3 The Federal Food, Drug and Cosmetics (FD&C) Act and The Fair Packaging and Labeling Act

The two Acts which form the cornerstone for food labelling regulations in the USA are the Federal Food, Drug and Cosmetics Act (FDA), and The Fair Packaging and

Labeling Act (FTC). These Acts contain general policies relevant to food labelling, which are used as a basis for regulations to lay down specific and detailed requirements. This section provides an overview of the general policies of the Acts, which underpin the labelling requirements to come.

2.3.1 The Federal Food, Drug and Cosmetics Act

Most of the legislation of interest to the food industry is that enforced by the FDA, in application of the Federal Food, Drug, and Cosmetic Act ("FD&C Act" or "The Act" is used by FDA; the abbreviations "FFDCA" and "FDCA" are also seen). The FD&C Act is a set of laws passed by Congress in 1938 that granted authority to FDA to oversee the safety of food, drugs, and cosmetics. The FD&C Act defines food as articles used for food or drink for man or other animals, chewing gum, and articles used for components of any such article.

Section 201(k) of the Act defines the term "label" to mean:

> *a display of written, printed, or graphic matter upon the immediate container of any article; and a statement, or other information appearing on the label shall not be considered to be complied with unless such word, statement, or other information also appears on the outside container or wrapper, if any there be, of the retail package of such article, or is easily legible through the outside container or wrapper (FDA, 1938).*

In other words, the "label" is on the package itself and, if an overwrap is used, the material printed on the label must be either on the overwrap or easily read through the overwrap.

However, it is interesting to note that Section 201(m) of the FD&C Act defines the term "labeling" to include more than just the label. "Labeling" means "all labels and other written, printed, or graphic matter

1. upon any article or any of its containers or wrappers, or
2. accompanying such article."

Promotional, educational, and other materials, such as in-store signage that may "accompany" a food, are considered labelling and subject to the same laws and regulations.

Section 403 of the Act defines misbranding of foods. A food is considered to be misbranded if:

- its labeling is false or misleading,
- it is offered for sale under the name of another food,
- it is an imitation of another food and does not bear the word "imitation" in proper type size and in the proper location on the label,
- the container is made, formed, or filled so as to be misleading,
- the label does not contain the name and address of the manufacturer, packer, or distributor, and an accurate statement of the net quantity of contents,
- required information is not present and in a form that is likely to be read and understood by consumers,

- it is represented as a standardized food and does not meet the requirements of the Standard of Identity for that food,
- the food is not a standardized food and the label does not contain the common or usual name of the food, if any, and
  - if fabricated from two or more ingredients, the label does not contain the common or usual name of each ingredient
  - it is a special dietary food and does not include information required by regulation,
  - if a food contains any artificial flavoring, artificial coloring, or chemical preservative and the labeling does not state that fact.

In addition, omitting certain "facts material" may cause a food to be misbranded according to Section 201(n) of the Act. Facts material is "important" information concerning the possible consequences that may result from the use of the food. This Section of the Act was the basis for requiring the labelling of sulfiting agents when used in food products, as well as the declaration of the colour Yellow #5 (Tartrazine in the EU). Certain other labelling violations, although technically considered to render a product "misbranded" (as opposed to "adulterated") may create a safety hazard and therefore be given higher priority by FDA. These include failure to declare a major food allergen and failure to label with a warning statement required by FDA.

2.3.2 The Fair Packaging and Labeling Act

The Fair Packaging and Labeling Act (FPLA) governs the labelling requirements for most consumer products (FTC, 1967).

The main provisions of interest are those relating to net quantity of contents declarations. The FPLA relates only to the net quantity of contents information on packages, goods, or commodities that are sold on the basis of weight or measure. Labelling requirements for packaged goods are applied to packages based on who will be the ultimate consumer. There are two classifications of products: "consumer" packages that are intended for retail sale. The other class is "non-consumer" packages that are intended for sale in wholesale trade, or business-to-business sales. The FPLA requirements are not applicable to all packaged goods, for those products under the jurisdiction of the USDA and TTB are outside the scope.

In addition to laws and regulations, some regulatory policies and practices are detailed in Compliance Policy Guides (CPG), which are issued and revised periodically by FDA.

2.4 The main labelling requirements according to the Federal Food, Drug and Cosmetics Act

We will now examine the detailed requirements promulgated to apply the FD&C Act, and which give the US food label some of its distinctive characteristics. This section also includes requirements for alcohol warning statements under the authority of other regulatory bodies, but which are also important to consider.

2.4.1 Field of vision

The following general requirements for the format and layout of labelling information are laid down in 21 CFR Part 101.

The principal display panel (PDP) is the part of the label most likely to be displayed, presented, shown, or examined under customary conditions of display for retail sale. It shall be large enough to accommodate the obligatory statements. The "area of the PDP" is measured as follows:

- For a rectangular package, one entire side – height × width;
- In the case of a cylindrical or nearly cylindrical container, 40% of the product of the height of the container and the circumference;
- In the case of any otherwise shaped container, 40% of the total surface of the container (under certain conditions).

The statement of identity and net quantity declaration must appear on the PDP, where the net quantity is in the bottom 30%. Both declarations must be in lines generally parallel to the base of the package.

The information panel (IP) is that part of the label immediately adjacent, and to the right of, the PDP as displayed to the consumer (with various exceptions). If this panel is not usable, due to package design (e.g. if covered by a fold), then the IP is the next label panel immediately to the right.

The list of ingredients, name and address of the manufacturer, packer or distributor, and nutrition labelling must appear on either the PDP or the IP. Information on the IP must appear in one place without interruption by illustrations, bar codes, or other non-mandatory text.

2.4.2 Language

In accordance with 21 CFR Part 101.15, all mandatory label requirements must appear in English. Where articles are distributed solely in the Commonwealth of Puerto Rico or in a territory where the predominant language is other than English, this may be substituted for English. However, if the label contains any statements in a foreign language, all words, statements, and other information required must appear in the foreign language or languages. FDA has stated it is preferable for foreign language labelling to be on an alternative PDP (FDA, 1989).

2.4.3 Intelligibility

All the information appearing on the PDP or IP must appear prominently and conspicuously in letters no less than one-sixteenth inch in height (1.6 mm), based on the lower-case letter "o." In addition, the height-to-width ratio of the lettering must not exceed 3:1. There are further type size requirements for the net quantity and nutrition labelling. The statement of identity on the PDP must be in bold face type, in a size reasonably related to the most prominent printed matter.

Various exemptions exist on the information on the IP for soft drinks in bottles, individual serving-size packages for foodservice, and where information is placed on a removable lid. Other exemptions are established for serving-size packages and individually wrapped confectionery pieces.

2.4.4 Statement of Identity

21 CFR 101.3 defines "statement of identity" (name of the food) as follows. This statement must be the name of the food as established by law, regulation, or policy or the accepted common or usual name of the food. If the food is covered by a Standard of Identity, the Standard establishes the common or usual name for the food that is required to be used. Standards of Identity are laid down for the following types of food, in 21 CFR Parts 131–169:

- Milk and cream
- Cheeses and related cheese products
- Frozen desserts
- Bakery products
- Cereal flours and related products
- Macaroni and noodle products
- Canned fruits
- Canned fruit juices
- Fruit butters, jellies, preserves, and related products
- Fruit pies
- Canned vegetables
- Vegetable juices
- Frozen vegetables
- Egg and egg products
- Fish and shellfish
- Cacao products
- Tree nuts and peanut products
- Beverages
- Margarine
- Sweeteners and table syrups
- Food dressings and flavorings

If there is neither a name established by law or regulation nor a common or usual name, an appropriately descriptive term or a fanciful name commonly used by the public for a food may be used, but a brand name cannot substitute. In addition to the standards laid down in regulations, there are also recognized Common or Usual Names for specific non-standardized foods under 21 CFR Part 102; for example, peanut spreads, beverages that contain fruit or vegetable juice, onion rings made from diced onion, potato chips made from dried potatoes.

If a new food resembling a traditional food and substitutes for the traditional food, it must be labelled as an imitation if the new food contains less protein or a lesser amount of any essential vitamin or mineral than its traditional counterpart. The same type size and prominence for the word "Imitation" as for the name of the product imitated must be used.

2.4.5 List of ingredients

For all foods composed of two or more ingredients, a list of ingredients must be declared, on the IP if not on the PDP, in descending order of predominance by weight, listed by common or usual name. It is common to see single-ingredient foods with an ingredient list, however, to emphasize there is nothing added. Requirements for ingredient declaration are laid down in 21 CFR 101.4.

Ingredients present in amounts of 2% or less may be placed at the end of the ingredients list using an appropriate statement, e.g. "Contains … percent or less of …," or "Less than … percent of …," with a threshold level of 2%, 1.5%, 1% or 0.5% may be used.

Generic names are permitted for certain ingredients, but the list of these ingredients is quite limited compared to that in the EU: skim milk or nonfat milk, milk, buttermilk, whey, cream, butterfat, eggs, egg whites, egg yolks.

2.4.5.1 Spices

Under 21 CFR 101.22, the term *spice* can be used when declaring any aromatic vegetable substance in the whole, broken, or ground form, except for those substances that have been traditionally regarded as foods, such as onions, garlic, and celery; whose significant function in food is seasoning rather than nutritional; that is true to name; and from which no portion of any volatile oil or other flavouring principle has been removed. It is interesting to note that paprika, turmeric, saffron, or other spices that are also colours, are to be declared as "spice and coloruing" unless declared by their common or usual name. Unlike in the EU, spices with a colouring effect are not excluded from the definition of "colour."

2.4.5.2 Fats and oils

Fat and/or oil ingredients are normally to be listed by their common or usual name in order of predominance when used in blends. However, those not present in the product may be listed if they may sometimes be used in the product. Such ingredients shall be identified by words indicating that they may not be present, such as "or," "and/or," "contains one or more of the following:," e.g., "vegetable oil shortening (contains one or more of the following: cottonseed oil, palm oil, soybean oil)." No fat or oil ingredient shall be listed unless actually present if the fats and/or oils constitute the predominant ingredient of the product. This flexibility is sometimes used by manufacturers of processed foods, when the constant use of a particular oil is not guaranteed by a supplier. The terms "hydrogenated," or "partially hydrogenated" are to be used in the ingredient name if applicable.

2.4.5.3 Added water

Added water must be declared in the appropriate position in the ingredients list, unless it is completely evaporated during processing.

2.4.5.4 Compound ingredients

An ingredient that contains two or more ingredients and has an established common or usual name must be listed in the ingredients list either:

- by declaring the established common or usual name of the ingredient followed by a parenthetical listing of all the ingredients contained therein in descending order of predominance, or
- by incorporating into the statement of ingredients, in descending order of predominance in the finished food, the common or usual name of every component of the ingredient without listing the ingredient itself.

2.4.5.5 QUID

QUID is not a requirement and is rarely seen, but percentages of ingredients *may* be shown in parentheses following the name of the ingredient and expressed in terms of percent by weight, to the nearest 1%.

2.4.5.6 Additives

The common or usual name is required for additives. E-numbers are neither used nor recognized, and may not be declared. Individual additives may have specific labelling requirements laid down in the regulations applicable to them. For example, laxative warning statements are laid down not in 21 CFR Part 101 but, in the case of sorbitol, in 21 CFR 184.1835: "The label and labeling of food whose reasonably foreseeable consumption may result in a daily ingestion of 50 g of sorbitol shall bear the statement: "Excess consumption may have a laxative effect." Also, 21 CFR 172.804 which regulates the use of aspartame states that the label of foods containing aspartame must declare: "PHENYLKETONURICS: CONTAINS PHENYLALANINE," either on the PDP or on the IP.

Additive category or function is generally not required, except in the case of chemical preservatives (or colours, which are not considered as "food additives"). However, the collective names "leavening," "dough conditioners," "yeast nutrients," "firming agents," followed by a listing of the specific additives in order of prominence. If the producer is unable to guarantee a constant pattern, any additives of these categories that may or may not be present can be declared using "or," "and/or," or "contains one or more of the following."

For preservatives, the usual name is declared and a separate description of its function, e.g. "preservative," "to retard spoilage," "a mold inhibitor," "to help protect flavour," or "to promote colour retention."

Incidental additives, as defined in the regulations, that are present in a food at insignificant levels and do not have any technical or functional effect in that food are not required to be declared in the listing of ingredients. Carry-over is not otherwise addressed for the purposes of labelling. However, any food that contains sulfiting agents that are functional or present in the finished food at detectable levels (defined as greater than ten parts per million) must be declared on the label. Sulfites are not listed among major food allergens; however, please see Section 2.5 on Allergen Labelling.

2.4.5.7 Colour additives

Colours that are exempt from certification may be declared as "Artificial Colour," "Artificial Colour Added," "Colour Added," or alternatively by declaring the name of the colour in the forms: "Coloured with beta-carotene," or "beta-carotene colour." In accordance with 21 CFR 73.100, cochineal extract or carmine is to be declared by listing its respective common or usual name, "cochineal extract" or "carmine." It should be noted that no colour additive in the USA is considered a "natural colour" and should not be declared as such. This is because it is not "natural" in the product to which it is added.

Colours that are subject to certification should be declared by their specific name, e.g. "FD&C Blue No. 2" or "Blue 2" (the terms "FD&C" and "No." are optional). The term "Lake" must be included in the declaration of the lake of a certified colour additive, e.g. "Blue 1 Lake." However, if FD&C Yellow No. 5 (Tartrazine in the EU) is used, it must be declared using its full name.

2.4.5.8 Flavourings

Natural flavouring and artificial flavouring may be declared as "natural flavour" and "artificial flavour" or any combination thereof. The term "flavour" or "flavouring" may also be used for flavours; however, when artificial flavours are used, this must be indicated. Pyroligneous acid or other artificial smoke flavours are declared as "artificial flavour" or "artificial smoke flavour."

Beyond the scope of declaration of flavourings used as ingredients, there are detailed requirements in 21 CFR Part 101.22 regarding label representations relating to primary recognizable flavours. These apply to use of words, vignettes (e.g. depiction of a fruit), or any other means. There are various options for declaration of such flavours depending on whether the characterizing flavour is natural or artificial and whether the flavour in the food is derived from the product whose flavour is simulated.

2.4.6 Name and address

Under 21 CFR 101, it is required to declare the name and address of manufacturer, packer or distributor, which must include the street address, city, State and ZIP code (or country and postcode if outside the US) or, if address given is in the US, the street address may be omitted if it is shown in a current city directory or telephone directory.

Where the food is not manufactured by the person named on the label, statements such as: "manufactured for …," "distributed by …," or similar must be used.

2.4.7 *Warning statements*

21 CFR 101.17 contains the following warning, notice, and safe handling statements. Further conditions and exceptions are laid down in that part.

- Self-pressurized containers: "WARNING —Avoid spraying in eyes. Contents under pressure. Do not puncture or incinerate. Do not store at temperature above 120°F. Keep out of reach of children."
- Self-pressurized containers with halocarbon or hydrocarbon propellants (in addition to the above): "WARNING – Use only as directed. Intentional misuse by deliberately concentrating and inhaling the contents can be harmful or fatal."
- Protein products in liquid, powdered, tablet, capsule, or similar forms that derive more than 50% of their total caloric value from either whole protein, protein hydrolysates, amino acid mixtures, or a combination of these, and that are represented for use in reducing weight: "WARNING: Very low calorie protein diets (below 400 Calories per day) may cause serious illness or death. Do Not Use for Weight Reduction in Such Diets Without Medical Supervision. Not for use by infants, children, or pregnant or nursing women."
- Protein products represented as part of a nutritionally-balanced diet plan providing 400 or more calories (kilocalories) per day, and the label specifies the diet plan in detail (or provides a description and information on how to obtain further details): "NOTICE: For weight reduction, use only as directed in the accompanying diet plan (the name and specific location in labeling of the diet plan may be included in this statement in place of 'accompanying diet plan'). Do not use in diets supplying less than 400 Calories per day without medical supervision."
- Protein products represented or intended for dietary (food) supplementation deriving more than 50% of their total caloric value from either whole protein, protein hydrolysates, amino acid mixtures, or a combination of these, represented specifically for purposes other than weight reduction: "NOTICE: Use this product as a food supplement only. Do not use for weight reduction."
- Dietary supplements containing iron or iron salts: "WARNING: Accidental overdose of iron-containing products is a leading cause of fatal poisoning in children under 6. Keep this product out of reach of children. In case of accidental overdose, call a doctor or poison control center immediately."
- Foods containing psyllium husk: (foods containing dry or incompletely hydrated psyllium husk, and which bear a health claim on the association between soluble fiber from psyllium husk and reduced risk of coronary heart disease): "NOTICE: This food should be eaten with at least a full glass of liquid. Eating this product without enough liquid may cause choking. Do not eat this product if you have difficulty in swallowing." However, a product in conventional food form may be exempt from this requirement if a viscous adhesive mass is not formed when the food is exposed to fluids.
- Juices that have not been specifically processed to prevent, reduce, or eliminate the presence of pathogens: "WARNING: This product has not been pasteurized and, therefore, may contain harmful bacteria that can cause serious illness in children, the elderly, and persons with weakened immune systems."
- Shell eggs: "SAFE HANDLING INSTRUCTIONS: To prevent illness from bacteria: keep eggs refrigerated, cook eggs until yolks are firm, and cook foods containing eggs thoroughly."

2.4.8 Alcoholic Beverage Warning Statement

Although not regulated by FDA, it is worth mentioning that TTB, in 27 CFR Part 16, lays down a requirement for an Alcoholic Beverage Health Warning Statement. This statement must appear on all alcohol beverages for sale or distribution in the US containing at least one-half percent (0.5%) alcohol by volume, intended for human consumption and bottled on or after 18 November 1989:

"**GOVERNMENT WARNING**: (1) According to the Surgeon General, women should not drink alcoholic beverages during pregnancy because of the risk of birth defects.

(2) Consumption of alcoholic beverages impairs your ability to drive a car or operate machinery, and may cause health problems."

This warning may appear on the front, back or side of the container. There are also specifications for the formatting and type size of the statement.

2.5 Allergen labelling

Allergen labelling, while an important focus in the US and EU, has also taken a divergent path. The "Food Allergen Labeling and Consumer Protection Act of 2004" ("FALCPA") (FDA, 2004) identifies the eight major foods or food groups that account for 90% of food allergies in the USA. The term "major food allergen" refers to *milk, eggs, fish, crustacean shellfish, tree nuts, peanuts, wheat and soybeans* and ingredients that contain protein derived from these specified foods and food groups. Highly refined oil derived from the specified foods and food groups, and any ingredient derived from such highly refined oils, are exempted from the definition of a major food allergen.

FALCPA gives two alternatives for the format of allergen declaration.

- "Contains," followed by the name of the food source from which the major food allergen is derived is to be declared immediately after or adjacent to the list of ingredients (in a type size no smaller than the type size used for the ingredient declaration). Other headings such as "Allergy Advice" are not to be used in connection with this statement.
- Alternatively, the name of the ingredient containing or from which the major food allergen is derived is followed by the name of the allergen in brackets, with the following exceptions:
  - the name of the food source is not required when the common or usual name of the ingredient uses the name of the food source from which the major food allergen is derived,
  - or the name of the food source from which the major food allergen is derived appears elsewhere in the ingredient list (unless that declared food ingredient is not a major food allergen).

Allergens may be declared both in the ingredients list and by a "Contains" statement, but each must be consistent (all allergens must then be declared by both methods). It is also interesting to note that, in contrast to the EU food information regulation, repetition is not required as the allergen does not have to be mentioned every time it occurs. The label scan in Figure 2.1 below illustrates this dual declaration, which is commonly seen.

The "name of the food source from which the major food allergen is derived" is defined in the Act as the name of the major food allergen as listed in the Act. Therefore "wheat" must be declared rather than "gluten."

Figure 2.1 An example of an ingredients list with allergens declared in the list of ingredients and highlighted in a "Contains" statement.

FDA has stated in their guidance "A Food Labeling Guide" (FDA, 2009) that "wheat" is considered to be any species in the genus *Triticum*. In addition, FDA identified the following as being "tree nuts" for the purposes of allergen labelling:

- Almond
- Ginkgo nut
- Beech nut
- Hickory nut
- Brazil nut
- Lichee nut
- Butternut
- Macadamia nut/Bush nut
- Cashew
- Pecan
- Chestnut (Chinese, American, European, Seguin)
- Pine nut/Pinon nut
- Chinquapin
- Pistachio
- Coconut
- Shea nut
- Filbert/hazelnut
- Walnut (English, Persian, Black, Japanese, California), Heartnut

Note that, unlike the EU, coconut and pine nut are also considered to be allergens.

Exempted ingredients are not listed in the Act, but FDA has established procedures by which manufacturers may notify or petition for allergen labelling exemptions for ingredients. At time of writing, FDA has granted exemptions for only two substances: soy lecithin, used as a processing aid, and ice-structuring protein from fish (FDA, 2013).

2.6 Net quantity of contents

For net quantity, 21 CFR 101.105 requires the following proportions to be observed, based upon the size of the PDP calculated as described under "Field of Vision," above.

Table 2.1 **Minimum type sizes for net contents declarations**

Minimum type size	Area of principal display panel
One-sixteenth inch (1.6 mm)	5 square inch (sq. in) (32 sq. cm) or less
1/8 inch (3.2 mm)	More than 5 sq. in (32 sq. cm) but not more than 25 sq. in (161 sq. cm)
3/16 in (4.8 mm)	More than 25 sq. in (161 sq. cm) but not more than 100 sq. in (645 sq. cm)
1/4 in (6.4 mm)	More than 100 sq. in (645 sq. cm) but not more than 400 sq. in (2580 sq. cm)
1/2 in (12.7 mm)	Over 400 sq. in (2580 sq. cm)

Minimum type sizes for declaration of net contents are laid down as shown in Table 2.1:

21 CFR 101.105 states: "The declaration shall appear as a distinct item on the PDP, shall be separated (by at least a space equal to the height of the lettering used in the declaration) from other printed label information appearing above or below the declaration and (by at least a space equal to twice the width of the letter "N" of the style of type used in the quantity of contents statement) from other printed label information appearing to the left or right of the declaration. The declaration shall appear in conspicuous and easily legible boldface print or type in distinct contrast (by typography, layout, colour, embossing, or molding) to other matter on the package." These regulations originate in the FPLA and not the FD&C Act, and are relevant to consumer goods generally.

2.7 Date marking and the uniform open dating regulations

There are no Federal requirements for the declaration of a durability indication. However, the US Department of Commerce National Institute of Standards of Technology (National Conference on Weights and Measures under the NIST) issued the Uniform Open Dating Regulation (NIST, 2014 edition), which provides guidance on this issue. It is not a Federal Regulation and has not been adopted by all states of the Union. In some cases the regulation has been adopted so that a "sell by" date is mandatory, but a "best if used by" date is optional.

The regulation defines different categories of food by risk:

- Perishable food": any food having a significant risk of spoilage, loss of value, or loss of palatability within 60 days of the date of packaging
- "Semi-perishable food": any food for which a significant risk of spoilage, loss of value, or loss of palatability occurs only after a minimum of 60 days, but within 6 months, after the date of packaging

Figure 2.2 An ink-jet-printed "Best When Used By" date.

- "Long shelf-life food": any food for which a significant risk of spoilage, loss of value, or loss of palatability does not occur sooner than 6 months after the date of packaging, including foods preserved by freezing, dehydrating, or being placed in a hermetically sealed container.

The "*Sell by*" date is to be used for perishable foods. It is expressed by month and day: the first three letters of the month followed by numeral indicating calendar day, or month represented numerically followed by a numeral indicating the calendar day, or day of week only for bakery products with a shelf-life of less than 7 days. Information may be separated by a point, dash, slash, or space.

"*Best if used by*" is to be declared for semi-perishable and long shelf-life foods, by: month and year, or the first three letters of the month followed by a numeral indicating year, and may include day of month provided this is placed prior to month (e.g. 30 JUN 13). The example in Figure 2.2 illustrates a jet-printed "Best When Used By" date:

The date is to be printed, stamped, etc., or otherwise shown on the package, label on the package, or attached tag in a manner that is easily legible and separate from other information or graphics so as to be clearly visible to a prospective purchaser. There are no field of vision requirements, but the date must not be superimposed on other required information or obscured by other information, graphics, or pricing. Signposting, stating or indicating the location of the date may also be used. The product pictured in Figure 2.2, above demonstrates that the presentation of this information can be inconsistent and also extends to the statements used, particularly in the case of non-perishable foods. In the example (see Figure 2.3) the consumer is only directed to look for the date with no location indicated, and, as is sometimes the case, the date is not clearly legible and is printed together with the production code.

2.8 Country of origin

Country of origin is not a requirement of FDA and is not addressed in 21 CFR. However, country of origin is a requirement of the U.S. Customs and Border

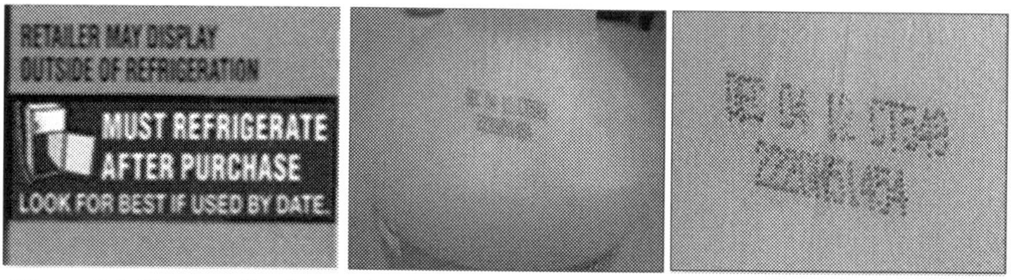

Figure 2.3 Non-specific signposting "Look for Best if Used by Date" on the BOP label of a beverage and the date printed on the bottle in a poorly legible format, together with production codes.

Protection, as authorized by the Tariff Act of 1930 and Customs' regulations (19 USC 1304(a) and 19 CFR Part 134). The country of origin must be conspicuously declared on the labels of imported food products (no location is specified).

"Country of origin" is defined as meaning the country of manufacture, production, or growth of any article of foreign origin entering the United States, to indicate to an ultimate purchaser in the United States the English name of the country of origin of the article, at the time of importation into the customs territory of the United States. The "ultimate purchaser" is generally the last person in the United States who will receive the article in the form in which it was imported.

2.9 Distance selling

FDA addressed the food labelling in terms of information on the Internet in a 2001 letter to the Washington Legal Foundation, in response to a petition requesting FDA cease to consider information on websites to be "labelling" (FDA, 2001). FDA responded that information disseminated over the Internet by, or on behalf of, a regulated company may meet the definition of labelling in Section 201(m) of the Act and be subject to its requirements. A website that promotes an FDA-regulated product, and allows consumers to purchase directly from the website, is likely to be "labelling," as it would be if the label for a product contained a statement referring the consumer to a specific website for additional information about a claim for the product. The websites, in these cases, are considered "written, printed, or graphic matter that supplements or explains the product and is designed for use in the distribution and sale of the product. FDA regularly issues warning letters, which often comment on information on company websites; of particular concern are claims made for dietary supplements. FDA stated in the 2001 letter that it had considered issuing guidance or regulations on the Internet presentation of food information, but that it felt this would not be able to keep pace with the fast-evolving nature of the Internet and that it would continue to evaluate web-based food information on a case-by-case basis.

2.10 The future of US food labels?

After a long period of relative quiet on the regulatory front, change may be on the cards in the near future. On 19 September 2013, two US Representatives and a Senator introduced a bill, H.R. 3147, entitled "The Food Labeling Modernization Act of 2013," noting that many of the general food labelling provisions of the FD&C Act were unchanged since 1938 and had not kept pace with the needs of the modern consumer (Congressional Fact Sheet, 19 September 2013).

The bill would mandate the Secretary of Health and Human Services (HHS) to issue comprehensive guidance for industry to clarify scientific support needed for a structure/function claim and legally empowering the Secretary to force companies to provide their substantiation documents.

Regarding general food labelling, the bill would also direct HHS to:

- require labels to disclose the amount of caffeine in a product if it exceeds 10 mg,
- Issue new format requirements for ingredient lists, including specific requirements for upper- and lower-case characters, font types, high-contrast between text and background, and use of bullet points for greater clarity.

At time of writing, the bill was referred to the House Committee on Energy and Commerce and an identical bill, introduced in the Senate in November 2013, was submitted to the Committee on Health, Education, Labor, and Pensions. The likelihood of its eventual adoption in its current form is uncertain, but this bill could herald a new era in informing US consumers about their food choices.

2.11 Sources of further information and advice

The following are useful resources for US food information recommended by the author:

Chapter IV of the Federal Food, Drug and Cosmetic Act, which covers food, can be accessed online at: http://www.fda.gov/RegulatoryInformation/Legislation/FederalFoodDrugandCosmeticActFDCAct/FDCActChapterIVFood/default.htm

The Electronic Code of Federal Regulations, although not an official source, is an updated, online version of the Code of Federal Regulations and can be browsed by Title: http://www.ecfr.gov/

FDA has a wealth of documents and guidance available on their "Labeling and Nutrition" page, including Nutrition Labelling, Front-of-Pack Labelling, Allergens and Claims: http://www.fda.gov/Food/IngredientsPackagingLabeling/LabelingNutrition/default.htm, and the Labeling & Nutrition Guidance Documents & Regulatory Information: Guidance for Industry page contains links to further, detailed guidance: http://www.fda.gov/Food/GuidanceRegulation/GuidanceDocumentsRegulatoryInformation/LabelingNutrition/default.htm

FDA's A Food Labeling Guide, Revised January 2013 is found on the above page but deserves particular mention. This guide, presented in a question-and-answer format by subject area is more a more user-friendly summarization of the major labelling requirements. www.fda.gov/FoodLabelingGuide

FDA's Compliance Policy Guides contain guidance on food labeling under Section 562. http://www.fda.gov/ICECI/ComplianceManuals/CompliancePolicyGuidanceManual/ucm119201.htm#SubChapter562

FDA's page on Topic-Specific Labelling Information contains details of areas which could not be fully explored in this Chapter: http://www.fda.gov/Food/GuidanceRegulation/GuidanceDocumentsRegulatoryInformation/LabelingNutrition/ucm2006864.htm

FDA, Questions and Answers: Gluten-Free Food Labeling Final Rule http://www.fda.gov/Food/GuidanceRegulation/GuidanceDocumentsRegulatoryInformation/Allergens/ucm362880.htm

FDA's Warning Letters can be a useful source of guidance and interpretation: http://www.fda.gov/iceci/enforcementactions/WarningLetters/default.htm

FDA's Information Pages regarding the two Proposed Rules to amend Nutrition Facts Labelling has a wealth of guidance, links to the publications, and examples of the modified Nutrition Facts panel: http://www.fda.gov/Food/GuidanceRegulation/GuidanceDocumentsRegulatoryInformation/LabelingNutrition/ucm385663.htm

National Organic Program: International Trade Policies: European Union. This page details the equivalency agreement between the USA and the EU: http://www.ams.usda.gov/NOPTradeEuropeanUnion

NIST Uniform Open Dating Regulation: National Institute of Standards and Technology (NIST Handbook No. 130, 2014 Ed.): http://www.nist.gov/pml/wmd/pubs/upload/17-section-IVe-14-h130-final.pdf

The Bill, Food Labeling Modernization Act of 2013 (H.R. 3147) can be accessed on the Library of Congress's Congress.gov website, at: http://beta.congress.gov/bill/113th-congress/house-bill/3147/text

TTB Ruling No. 2013–2, Voluntary Nutrient Content Statements in the Labeling and Advertising of Wines, Distilled Spirits, and Malt Beverages http://www.ttb.gov/rulings/2013-2.pdf

USDA's FSIS page on Labelling and Label Approval give an extensive overview on policies, procedures and guidance relevant to USDA-regulated products. The "Food Standards and Labeling Policy Book", which contains USDA"s "Natural" policy definition is available under "Labeling Policies": http://www.fsis.usda.gov/wps/portal/fsis/topics/regulatory-compliance/labeling

References

Congressional Fact Sheet, The Food Labeling Modernization Act of 2013, Congress of the United States, 19.9.2013.

FDA (2013), Investigations Operations Manual, Exhibit 3-1, p. 99.

FDA Basics for Industry, FDA website, "About FDA." Available from http://www.fda.gov/AboutFDA/WhatWeDo/default.htm

FDA (1938), Food, Drug and Cosmetic Act. P.L. 75–717, 52 Stat. 1040.

FDA (2004), Food Allergen Labeling and Consumer Protection Act of 2004 (Public Law 108–282), 2 August 2004.

FDA, Compliance Policy Guide 562.750, Rev. 2/1/1989, Manual of Compliance Policy Guides.

FDA, Guidance for Industry: Voluntary Labeling Indicating Whether Foods Have or Have Not Been Developed Using Bioengineering; Draft Guidance, January 2001.

FDA, Soy Lecithin Agency Response, Docket No. FDA-FL-0471, 22 February 22 2013.

FTC (1914), Federal Trade Commission Act of 1914 as amended, 15 U.S.C §§ 41–58.

FTC (1967), Fair Packaging and Labeling Act. 15 U.S.C. §§ 1451–1461.

National Institute of Standards and Technology (NIST), Uniform Open Dating Regulation, NIST Handbook No. 130.

State of California, California menu labeling law (SB 1420).

The Code of Federal Regulations, Titles 21, 9, 27, 16, 19. US Government Printing Office, 2013. Available online at: http://www.ecfr.gov (Accessed 27 September 2013).

Enforcement of food and beverage labelling legislation: enforcement bodies and relevant legislation in the UK

3

I.N. Hancock
Trading Standards Service, Dorset County Council, Dorchester, UK

3.1 Introduction

A food business owner, considering the labelling requirements for UK food and beverage products, or indeed anyone researching legal requirements and enforcement in this field, may be quite understandably intimidated by the jigsaw of arrangements that exist in terms of local enforcement and how that fits into a national framework to maintain food standards. Given the typical representation of trading standards, environmental health or other regulatory officers in the media, and continued outdated references to untypical incidents in other public forums, often politically motivated, those new to engaging with food law regulation may be apprehensive.

This and the following chapter seek to outline the jigsaw, and explain who enforces labelling requirements and how, including the context in which those services are provided. The chapters will hopefully provide some detail around the reality of local enforcement in the context of the national framework and reassurance as to the support and advice that a business can obtain from a local enforcement service, whether through trading standards or environmental health professionals.

This chapter focuses on the range of bodies involved in enforcing legislation related to food and beverage labelling. The legal framework these bodies must enforce is outlined, and includes consideration of key legal requirements affecting food description beyond the more obvious food labelling controls. The following chapter then describes the policies that have been devised to guide those responsible for enforcement. Enforcement bodies themselves need to ensure they meet national expectations to ensure this food enforcement jigsaw maintains a consistent approach which is supportive of businesses. Having absorbed the information in these chapters, it is hoped the reader will be able to engage with the appropriate local enforcer with some confidence and with a sound idea as to what to anticipate.

Advances in Food and Beverage Labelling. http://dx.doi.org/10.1533/9781782420934.1.35

3.2 The enforcement jigsaw: local government food authorities

Local authorities throughout the UK provide the services that enforce food and beverage labelling provisions. They do so as a result of the statutory duty to so in the keystone of this area of law, The Food Safety Act 1990, of which the relevant portion of Section 5 is reproduced below in its amended current form:

> *5 Food authorities and authorised officers.*
>
> *(1) Subject to subsections (3) and (4) below, the food authorities in England are –*
>
> *(a) as respects each London borough, district or non-metropolitan county, the Council of that borough, district or county;*
>
> *(b) as respects the City of London (including the Temples), the Common Council;*
>
> *(c) as respects the Inner Temple or the Middle Temple, the appropriate Treasurer;*
>
> *(d) as respects the Isles of Scilly, the Council of the Isles of Scilly.*
>
> *(1A) Subject to subsection (3)(a) and (b) below, the food authorities in Wales are, as respects each county or county borough, the Council of that county or county borough.*
>
> *(2) Subject to subsection (3)(a) below, the food authorities in Scotland are the councils constituted under Section 2 of the Local Government etc. (Scotland) Act 1994.*

There are further provisions specifically giving Port Heath Authorities similar responsibilities in port areas.

The Food Labelling Regulations 1996 are the other key piece of legislation governing labelling, and again the enforcement responsibility is specifically given to these "food authorities." These regulations are being replaced by the Food Information Regulations 2013 from the end of 2014, with many provisions coming into force subsequently. Other food labelling and composition regulations also refer back to the Food Safety Act to allocate responsibility in the same way to local authorities. The relevant legislative controls that are enforced by trading standards are outlined later in this chapter.

An officer enforcing the Act, or any piece of food legislation, will be authorised specifically by the local authority in writing. Also national guidelines from the Food Standards Agency (FSA) the Framework Agreement on Official Feed and Food Controls by Local Authorities, together with the Food Law Code of Practice, specify the qualification, competency and on-going training requirements for officers. An officer dealing with food labelling matters for a local authority should thereby be sufficiently skilled and knowledgeable to meet the needs of both business and the enforcement authority. There are broadly three types of officer involved in carrying out food work under the Food Safety Act and subordinate legislation:

Trading Standards Officer: with one of a range of qualifications in trading standards law, including food standards as required in the Food Law Code of Practice (Revised 2014). The officer may specialise in food standards, but is more likely to enforce a range of legislation. Not all "trading standards officers" are qualified in food standards law. Historically, and by statute, it has been a qualification in weights and measures law and practice that has defined a trading standards officer.

Environmental Health Officer: with a degree in Environmental Health from a university recognised by the Chartered Institute of Environmental Health and having undertaken an experiential learning portfolio. Likely to be more specialised in food hygiene requirements.

Public Analyst (and Port Health Analyst): possessing a Masters in Chemical Analysis awarded by the Royal Society of Chemistry, and having 3 years experience examining food in a certified, university or government laboratory. It is to these analysts that formal food samples are submitted under the Food Safety (Sampling and Qualifications) Regulations 2013.

The statutory duty makes it clear that enforcement responsibility lies with local government.

Businesses, especially those new to the UK enforcement scene, or new small-and medium-sized businesses, may initially be uncertain as to which local authority to turn to or expect to be dealing with, and in some areas the division between food hygiene, food composition and standards work may be unclear until regular contact is established. Unless a business has become embroiled in a major food incident, such as a national food alert, recall or food scare, it will be a local authority that will take the lead.

Throughout England the nature of local authorities varies, with just over 400 in total. In the 28 London Boroughs (including Westminster, Camden and Croydon) and the 55 unitary councils across England (examples include counties such as Cornwall and Wiltshire; clusters of metropolitan districts such as Birmingham, Manchester, Liverpool and their surrounding authorities and some major towns such as Poole and Bournemouth and the six authorities of Berkshire), there is one local authority in each area responsible for all food enforcement work – food hygiene and standards, composition and labelling. This work may be done by combined environmental health and trading standards teams. Alternatively, these professionally distinct disciplines, with separate qualification routes, may exist as separate teams, although liaising closely. In some of these unitary authority services, environmental health officers will be responsible for food labelling and composition enforcement as well as food hygiene, but you would be very unlikely to find trading standards officers engaged in food hygiene work, given the different professional training routes.

This local authority framework, a seemingly ever changing jigsaw redesigned by politicians through time, has for decades been subject to specific local arrangements between groups of local authorities that have joined together in joint service partnerships to provide services. As pressures to cut budgets continue, these arrangements are continually being reconsidered. One of the longest standing joint service arrangements is that at West Yorkshire Trading Standards Service, which has been serving the five metropolitan district councils in the area, including Leeds, since the break-up of greater metropolitan councils in 1986. More recently, Devon and Somerset County Councils have agreed to provide a joint trading standards service, while in Worcestershire a joint regulatory service has been established embracing what was previously six district council environmental health departments and the County trading standards service.

Local authorities can contract out services, as the Scilly Isles have done to enable Cornwall Council work in that area. A few arrangements have also been developed whereby the relevant regulatory service grouping has been contracted out to a private organisation, such as has been seen in North Tyneside, and also in the London Borough of Barnet where environmental health and trading standards services are operated on the Council's behalf by Capita Symonds. A food business may be unaware of such arrangements, and the services should not differ noticeably for those affected by regulation.

In county council areas where there remains a "two tier" structure, there is local responsibility for labelling, food composition and standards, including chemical contamination, while district councils provide food hygiene services through their environmental health services.

In Wales there are 22 unitary councils where food labelling responsibilities follow a similar pattern to English unitary councils but looking ahead a move to fewer and larger authorities can be expected. In the 32 unitary local authorities in Scotland, food labelling and other food regulatory functions can be found in the environmental health services. In Northern Ireland, all food enforcement work is done by environmental health officers within their 26 local council areas.

Any food business is required to be registered with an environmental health service, the food safety authority. However, for advice, guidance or discussion on areas of concern regarding food and beverage labelling, standards and composition, a business needs to seek out "an officer who enforces food labelling provisions" amid this national jigsaw of regulatory services.

3.3 Wider responsibilities of local food regulators

Whatever type of local authority fulfils the food labelling and standards role, and whether enforcement is by trading standards officers or environmental health officers, food work is one part of a wider portfolio of enforcement responsibilities competing for attention, resources, officer time and priority.

Food labelling, and other food standards work, may rest with a more specialist food team, but this is increasingly rare. In environmental health services, food hygiene work remains a high priority, with duties around health and safety at work, housing standards, pest control, littering, noise nuisance, pollution control and licensing functions often competing for resource.

Trading standards services, the most common home for the food labelling and standards role, have statutory enforcement roles to control unfair trading, including laws on: weights and measures; misleading claims; price marking; other fair trading issues; product safety; and animal health, disease control, welfare and feeding stuffs. This more general consumer law enforcement approach is not a model followed in many countries. Indeed the recent review of food enforcement in the UK, via the Food

Standards Agency (2013), considered greater centralisation of the food enforcement role but did not subsequently seek to move away from reliance on local authority enforcement. The more common European model is to have a specialist food enforcement agency.

A trading standards officer considering food labelling and composition will have a statutory interest in other areas of compliance which can assist in ideally addressing weight marking issues, and weight and measures packing requirements alongside food labelling concerns. One frustration can be that officers with the necessary competence and legal training in food law may not also have the competence and required qualifications and authorisation to deal with weights and measures issues. This will be particularly true where the trading standards service does not enforce food law.

Experience and expertise will vary. Local services will aim to ensure officers handling food standards and labelling issues are trained and qualified in relevant food law to the standards set nationally by the FSA, but it is the experience of dealing with particular food businesses and particular problems that brings real value. The best situation is where officers have experience and are dealing with key food business in their area of experience and competence. Also, there are occasions when the labelling, composition and associated marketing methods and description issues can be in conflict with trading standards law not specifically designed to address food, and again the overview of trading standards law can prove helpful.

3.4 National and regional coordination of local authority enforcement

Following the 2011 review of the consumer landscape by the Government (Department for Business, Innovation and Skills (BIS)), the National Trading Standards Board (NTSB) was formed, and has been operating fully from April 2012. Chaired by Lord Toby Harris, it comprises a group of senior and experienced heads of local authority trading standards services, representing all English regions and Wales. This reorganisation of responsibilities at national level was focused on fair trading (BIS areas of responsibility) and did not include food work, but the NTSB is clearly a valuable mechanism for achieving national coordination of enforcement. It remains to be seen to what extent its duties extend into food law, but this route has already been used recently to help address specific needs around imported food sampling and animal feed sampling to meet specific FSA needs. In 2012–13 NTSB allocated £13.7 million of Government funding to national and regional projects and major enforcement cases.

Previous technical and policy support for areas including food law has been lost since Local Government Regulation (LGR) was cut from the Local Government Association's services. The NTSB is now, with its small central team, supporting the Board, providing a professional support and network mechanism which is assisting local authority enforcement response to a wide range of fair trading issues and food law would be a natural extension of that. "Trading Standards Scotland" is now a mechanism seeking to coordinate local authority food enforcement work north of the border, where environmental health officers largely deal with all food enforcement.

Regional coordination of trading standards work, including food composition and labelling, is more established in England and Wales. Informal arrangements developed when funding in 2006–7 started to facilitate regional enforcement teams ("Scambuster" teams) which aimed to reduce consumer detriment by tackling rogue traders and scams. These regional coordination groups have helped coordinate food labelling projects and imported food sampling across group of authorities. The Trading Standards Partnership South West is one such arrangement, and examples of their work can be found on their website, including a toolkit for Chinese and Indian takeaway food businesses. The FSA has also linked into these regional groups via representatives on key liaison groups.

These regional groups, as with the NTSB and the FSA, are not the first point of contact though for any food labelling, or other trading standards issues, that business have: the local authority enforcers are the starting point, whether in trading standards or environmental health teams.

The FSA has been given the overall responsibility in the UK for food safety and hygiene, ensuring the food law of the European Union is complied with. The FSA is an independent government department given its statutory responsibility through the Food Safety Act 1990. While it may give guidance, undertake study, surveys and research and will comment on food issues nationally, directing local authorities on enforcement and providing specialist support, it does not deal directly in enforcing food labelling legislation. Only in very specialist circumstances would a business be expected to be in direct contact with the FSA rather than with the local authority enforcer. Difficulties with technical interpretation and dealings with other EU member states may be an exception, but the usual point of contact will be a local authority.

The FSA's stated vision is for "safer food for the nation." Its aims are to ensure UK produced food and imported food is safe to eat, that consumers have the information they need to make informed choices on where and what they eat, and that enforcement by local authorities is risk-based and focused on improving public health. They use the Framework Agreement on Official Feed and Food Controls (FSA April 2010) to provide guidance to local authorities on how they enforce food labelling and other food law provisions. The Agency can and does audit services and puts in place action plans for improvements where appropriate, with the ability to step in with alternative enforcement arrangements if services are failing to meet standards.

The FSA's labelling work has seen extensive consultation and consideration on front-of-pack nutritional labelling to encourage consumers to make healthier, or at least better informed, eating choices. The overall need has been to establish one recognisable, meaningful and easily understood pack front labelling option, albeit only with the larger companies involved. The horsemeat substitution scandal is an area where leadership was initially wanting, but coordination of the response to that has been a recent major task. Work to get food producers to voluntarily remove certain colourings linked to hyperactivity in children is an example of changes sought through non-statutory negotiation.

Early in 2013, a review into official food controls concluded that while the existing system, including local authority enforcement, was under pressure, these authorities could still deliver the service with better FSA collaboration being recognised as needed.

To add to the food law jigsaw that the UK has produced for itself, following changes in 2010, the Government injected a further source of confusion for businesses, consumers and enforcers, by giving the Government Department for Environment, Food and Rural Affairs (Defra) responsibility in England for food labelling and policy. The FSA continued to lead on "the food safety aspects" of food labelling and continued to liaise with food authorities in relation to food labelling law enforcement. Meanwhile in Scotland, Wales and Northern Ireland, the FSA remained in overall control of food responsibilities. It is perhaps odd, given that the horsemeat scandal was expressly said not to be a safety issue, that the FSA did not leave it for Defra to deal with at national level.

Fragmentation of responsibilities also saw the Goverment Department of Health take on nutritional labelling policy duties, as did the Welsh Assembly, with the FSA retaining this in Scotland and Northern Ireland.

Overall responsibility for the activity of local enforcement authorities and officers remained with the FSA.

A further consideration with this jigsaw is that certain foods and beverages may stray into the definition of medicines, thereby attracting the attention of the UK Medicines and Healthcare Products Regulatory Agency (MRHA), and overall policy in this area then becomes a Department of Health concern at central government level. Any substance or combination of substances having properties for treating or preventing human disease would, given the definition in EU Directive 2001/83/EC, be the subject of MRHA regulation. An especially relevant reference point is their Borderline Section, and further guidance can be found on the MRHA website.

In relation to weights and measures marking of food and beverages, at Government level, the Department for Business Innovation and Skills carries the national strategic responsibility, with the National Metrological Office (NMO), an executive agency of government, delivering this.

Aside from legislation, claims made in advertisements in any media may be subject to challenge by the Advertising Standards Authority (ASA), an independent regulator. The ASA may seek to challenge claims, and the presentation of food products, in any form of advertisement including newspapers, radio and television, text messaging and on the Internet. The ASA can ask for advertisers to justify claims, are not constrained by the legal procedures associated with enforcement, and their adjudications require less severe standards of proof. Failure to respond to challenges is likely to lead to complaints being upheld. The sanctions they use are the publication of adjudications and their ability to have publishers withdraw advertisements if the material fails to meet their code of practice. Complaint decisions are also publicised and attract negative media attention for businesses. Further detail on their codes and sanctions can be found on the ASA website.

3.5 National arrangements for consumer complaints about food and beverage labelling

Complaints and information about food labelling and composition are clearly one of the ways in which enforcement is prompted once a product is on the market, so it is worth highlighting what the usual routes for such complaints are.

The main route for consumer complaints about food labelling or any trading standards matter is now via the national Citizens Advice consumer service. Since April 2012, Citizens Advice have provided a national consumer helpline (03454 04 05 06). This is funded by central government to provide first line consumer advice and also to act as the first point of telephone contact for trading standards authorities in England, Scotland and Wales. It is distinct from local Citizens Advice Bureaux (CAB), the independent advice services operated locally under the Citizens Advice umbrella. This approach was built on a model of Consumer Direct, the previous national helpline operated by a range of regional contact centres.

It is usual for any consumer seeking to make a complaint or raise any concern about food composition or labelling to a trading standards authority directly, to have their call forwarded, automatically or manually, to the Citizens Advice consumer service. With significant resource constraints, local authorities are keen to channel consumer complaints through Citizens Advice with many not having a public number for consumers to call to ensure this happens and that costs are minimised. Complaints can also be made via the Citizens Advice website and by email.

Citizens Advice advisers, based in five call centres, one each in Wales and Scotland and three in England, provide information and guidance to the consumer, and appropriate cases are referred overnight, or immediately if urgent, to trading standards authorities. Referrals are by secure email from the Citizens Advice national database, and these may be marked as requiring action, with or without a consumer call back, or be simply notifications for information.

Businesses, including those aggrieved by unfair competition and competitors' labelling or products, are also a significant source of complaints that may lead to enforcement officer intervention. Most local trading standards authorities provide telephone advice for businesses on trading standards law, and this can be the most effective route into food law enforcement locally. Alternatively, e-mail or web-based contact will solicit a response, along with traditional letters or faxes, to a trading standards authority, or general local authority contact points.

Many consumers and business are understandably confused as to who to contact with food standards and labelling issues. Some are referred or signposted via the FSA, some from other local authority services, as well as the common referrals between trading standards or other food enforcing authorities.

Having explained this elaborate jigsaw of pieces it should hopefully be understood that, for businesses, local authority enforcement teams are the place to start, while consumers should call the Citizens Advice consumer service or contact via their website. A helpful route to the relevant local authority enforcer for food law can be found in the Enforcement and Regulation section of the FSA website, where a business can can be searched by postcode and locate relevant trading standards and environmental health teams.

3.6 Outline of relevant legislation relating to food and beverage labelling

This section summarises the key legislation that is enforced in relation to the labelling of food and beverages. The emphasis here is on the most common areas where enforcement action is taken.

3.6.1 Food Labelling Regulations 1996 and Food Information Regulations 2013

The Food Labelling Regulations 1996 (FLR) have been the principle source of detail in relation to food and beverage labelling. These regulations will be replaced from December 2014 by the provisions of the Food Information Regulations 2013, which are required for compliance with Regulation EU No 1169/2011.

The most common breach of FLRs that has led to prosecutions has been the sale of food passed its "use by" date (Regulation 44). A significant Supreme Court Appeal considered this requirement in the case of Torfaen County Borough Council vs Douglas Willis Ltd [2013] UKSC 59. That judgement underlined that where food was on sale beyond its "use by" date the prosecution simply had to show that it was marked with such a date and that the date had passed when the food was sold. There was no need to show that the food, in this case frozen meat, was in a highly perishable condition at the time of the offence. The Supreme Court explained that to read in further requirements as to the state or perishability of the food would be to weaken controls and reduce the ability of consumers to know how long the meat in question had been frozen. Formal action in relation to the sale of out-of-date food had been effectively curtailed while this judgement was awaited.

In practice, the newer Food Information Regulations do not significantly change the basic need for food to be honestly described and presented with pre-packed goods having a true name and a list of ingredients in descending order of weight. This gives consumers clarity about contents to make comparisons and choices if they so desire. Specific quantitive indications need to be added if any ingredients are given particular emphasis to help characterise the product.

Compliant labelling will indicate: how the product is to be handled, stored or prepared; who the manufacturer, packer or importer is; its origin, if failure to mark would mislead; any specific allergenic ingredients; if it is irradiated or contains genetically modified material; if it contains aspartame, high levels of caffeine, sweeteners; if it is packed in a protective gas; and alcohol content if greater than 1.2%, and include a net quantity marking. Full details can be seen on the GOV.UK and FSA websites.

Notable changes which have been attracting the attention of local regulators include the requirement to label and alert consumers to the presence of the common listed allergens in non-pre-packed foods from December 2014, and mandatory nutritional labelling on pre-packed food and non-alcoholic drinks from December 2016, together with a variety of additional matters of detail. Previously, nutritional information was only required where particular claims emphasised that a food was higher or lower in a particular nutritional characteristic.

Business to business supplies are covered by labelling requirements, so it is required that sufficient information is given to enable any supplier to be able to provide the mandatory information to their customers.

There are a range of particular exemptions that were in the FLRs and not under the new controls, which may prove problematic for some suppliers.

From December 2014, the quantity labelling of pre-packed food and drink will also need to meet Regulation EU No 1169/2011 on the "Provision of Food Information to Consumers." The UK is revoking its existing national legislation on quantity marking, along with various exemptions that exist. Quantity labelling will need to be visible, easy to read and durable. Prior to this change, the Weights and Measures (Packaged Goods) Regulations 2006 outline the related requirements. There are also existing requirements in relation to the sale of loose products by quantity, and other considerations under weights and measures legislation which need to be taken account of, and your local trading standards office can advise on these.

There is also a range of compositional and labelling regulations dealing with particular foods and beverages, also enforced locally, including spirit drinks, olive oil and genetically modified organisms, for example, and indeed some that are enforced by other national bodies, such as those dealing with wine. These can be explored more fully on the central government ".Gov" website.

3.6.2 Food Safety Act 1990

The key general provisions of the Food Safety Act 1990 are often also used to regulate misleading or incomplete labelling or information provided with products. Section 14 requires that consumers are only supplied with food and drink that is of the "nature, substance and quality" demanded by consumers. Much has been made in legal case law, and indeed a whole chapter could be written, regarding which term applies to which problem, but enforcers will be able to identify an offence if the food or drink supplied is not of the type, variety, or quality that one might expect given the standards, industry practice or descriptions used, or if there is anything present that should not be. Section 15 of the Act deals generally with the misleading presentation of food:

> *Section 14: Selling food not of the nature or substance or quality demanded.*
>
> *(1) Any person who sells to the purchaser's prejudice any food which is not of the nature or substance or quality demanded by the purchaser shall be guilty of an offence.*
>
> *(2) In subsection (1) above the reference to sale shall be construed as a reference to sale for human consumption; and in proceedings under that subsection it shall*

not be a defence that the purchaser was not prejudiced because he bought for analysis or examination.

Section 15: Falsely describing or presenting food.

(1) Any person who gives with any food sold by him, or displays with any food offered or exposed by him for sale or in his possession for the purpose of sale, a label, whether or not attached to or printed on the wrapper or container, which –

(a) falsely describes the food; or

(b) is likely to mislead as to the nature or substance or quality of the food, shall be guilty of an offence.

(2) Any person who publishes, or is a party to the publication of, an advertisement (not being such a label given or displayed by him as mentioned in subsection (1) above) which –

(a) falsely describes any food; or

(b) is likely to mislead as to the nature or substance or quality of any food, shall be guilty of an offence.

(3) Any person who sells, or offers or exposes for sale, or has in his possession for the purpose of sale, any food the presentation of which is likely to mislead as to the nature or substance or quality of the food shall be guilty of an offence.

(4) In proceedings for an offence under subsection (1) or (2) above, the fact that a label or advertisement in respect of which the offence is alleged to have been committed contained an accurate statement of the composition of the food shall not preclude the court from finding that the offence was committed.

(5) In this section references to sale shall be construed as references to sale for human consumption.

3.6.3 General Food Regulations 2004

The General Food Regulations 2004 makes it an offence to fail to comply with the General Food Law Regulations (EC) 178/2002. Again, a general approach means that labelling, advertising and presentation must not mislead consumers. Food businesses throughout the supply chain must be able to identify their source of supply, which is often a problem where retailers have bought non-compliant stock from informal sources. Provisions also deal with the need to withdraw unsafe food from the market, which could include where allergens are not marked clearly.

3.6.4 Fraud Act 2006: key sections

Trading Standards authorities have developed their use of the Fraud Act 2006 considerably in recent years and "food fraud" is a particular focus for the FSA, coordinating and supporting investigations. Where offences regarding the description, labelling and composition of food and beverages can be shown to be "dishonest" then fraud offences may have been committed by an individual. The key sections are set out below. The test of what is "dishonest" is set out in the appeal case of R v. Ghosh

[1982] EWCA Crim 2. Essentially, if a jury would think on the evidence heard that the practice was "dishonest," then it is. Fraud represents the most serious offence that a food supplier could be convicted of, and it is these offences that led to police involvement in investigations into the undeclared horse meat substitution scandal in 2013 and could be used to tackle any deliberate attempts to mislead or benefit from misdescription.

The maximum penalties are an unlimited fine or 10 years imprisonment at Crown Court level or, if dealt with by a magistrates' court, £5000 and/or 12 months imprisonment. For a food business to be associated with a conviction for fraud clearly generates problems in a different league, especially where the fraud involves senior staff or those that run the business:

1 Fraud

(1) A person is guilty of fraud if he is in breach of any of the sections listed in subsection (2) (which provide for different ways of committing the offence).

(2) The sections are –

(a) Section 2 (fraud by false representation),

(b) Section 3 (fraud by failing to disclose information), and

(c) Section 4 (fraud by abuse of position).

(3) A person who is guilty of fraud is liable –

(a) on summary conviction, to imprisonment for a term not exceeding 12 months or to a fine not exceeding the statutory maximum (or to both);

(b) on conviction on indictment, to imprisonment for a term not exceeding 10 years or to a fine (or to both).

(4) Subsection (3)(a) applies in relation to Northern Ireland as if the reference to 12 months were a reference to 6 months.2 Fraud by false representation

(1) A person is in breach of this section if he –

(a) dishonestly makes a false representation, and

(b) intends, by making the representation –

(i) to make a gain for himself or another, or

(ii) to cause loss to another or to expose another to a risk of loss.

(2) A representation is false if –

(a) it is untrue or misleading, and

(b) the person making it knows that it is, or might be, untrue or misleading.

(3) "Representation" means any representation as to fact or law, including a representation as to the state of mind of –

(a) the person making the representation, or

(b) any other person.

(4) A representation may be express or implied.

(5) For the purposes of this section a representation may be regarded as made if it (or anything implying it) is submitted in any form to any system or device

designed to receive, convey or respond to communications (with or without human intervention).

Section 3, which deals with dishonestly omitting information which is required to be given, Sections 6 and 7 which deal with article used to commit fraud, and Sections 9 and 10, which deals with individuals and companies operating fraudulent businesses, may also become relevant at the more criminal extremes.

3.6.5 Trade Marks Act 1994

The Trade Marks Act 1994 has been regularly used in recent years by local authority enforcers in relation to food, and particularly drink, supplies, most commonly in relation to the supply of counterfeit copies of branded spirit drinks. In some cases counterfeit "duty paid" stamps have also been identified. These offences have commonly led to prosecution of retailers, and the scenario is often that a small retail business has bought stock from an untraceable source at a knock-down price, making tracing the sources of crimes more challenging.

Section 94 of the Act 1994 makes it an offence to supply, offer, etc., any goods bearing a registered trade mark without authority from the rightful trademark owner. Forfeiture Orders may also be obtained, with or without a prosecution, for non-compliant stock (Section 97). These offences attract a maximum sentence of 6 months imprisonment and/or fines of up to £5000 in a magistrates' court and a substantial maximum of 10 years imprisonment and/or an unlimited fine. Further, they are considered as "lifestyle offences" under the Proceeds of Crime Act, which can trigger investigations into the finances of offenders and the need for offenders to account for income earned over a previous 6 year period or face significant financial confiscation. Counterfeit food products are certainly a problem for suppliers to avoid, and record keeping to trace supplies is particularly important with retailers' use of unverified sources of supply, potentially being a very costly mistake.

3.6.6 Consumer Protection from Unfair Trading Regulations 2008

The overall misdescription of food or the misleading of buyers may also give rise to consideration of the Consumer Protection from Unfair Trading Regulations 2008 (CPUT Regs), enforced by trading standards officers and very general in their application. While the Food Safety Act's purpose is primarily to provide food standards and quality controls, these regulations deal with the consumers' economic interests: where they are buying a product that they would not have bought had they known the truth about it. The CPUT Regs deal with commercial practices, including any form of representation which is misleading, or misleading by omission of key information. These are referred to as misleading actions and misleading omissions. The Regulations also prohibit a range of 31 specific practices that may be considered unfair commercial practices, including misuse of approval marks, for example, but most of which are not particularly relevant to food supplies.

The Business Protection from Misleading Marketing Regulations 2008 (BPR), introduced at the same time as the CPUT Regs, provide protection for businesses in their dealings with other business. They have similar penalties and prohibit misleading advertising of any sort, even oral statements, being given to businesses.

So from the very specific product-based regulations to broad general requirements for honest, accurate information to be provided, there is a wide range of regulatory control available to local enforcers to use where consumers or others are misled, with a range of sanctions that can flow from them.

3.7 Future trends

The changing consumer protection landscape will continue to see improved coordination of local consumer protection enforcement generally at national and regional level, and the opportunity is there for this to embrace food work to a greater extent. This is against a background of public sector cuts that will see further reductions to local food authorities, reducing their ability both to advise and support business and indeed take formal action. The new Regulators' Code will give further reason to focus the reduced enforcement resources on those high risk activities and businesses.

The jigsaw of food labelling enforcement will continue to change in shape as local arrangements to merge services continue, based on budget saving opportunities and local politics rather than any attention to effective local food enforcement. Finances may also see authorities seeking to recover costs for business advice, whether that is through the Primary Authority Scheme or other local charging arrangements.

The main challenge for businesses and regulators on labelling is compliance with the Food Information Regulations, in particular in those areas where previous labelling exemptions have been removed and where additional requirements are introduced. The drive to ensure that the presence of known allergens is clear on pre-packed food will continue, with it already being a major reason for food recalls and alerts, and with the extension to drawing attention to allergens in non-pre-packed food this will give the known allergens a high profile.

After 2016, nutritional labelling will be more widespread as it will no longer be voluntary and businesses may find the accuracy of this process more challenging in some areas, and it will be an added burden for some, or perhaps an opportunity to distinguish their products on nutritional grounds.

The detailed requirements for food and beverage labelling, and enhanced mandatory nutritional labelling, will continue to attempt to ensure consumers are protected, expected standards are maintained, and allow informed and hopefully healthier eating choices to be made. However, after a period when inadequacies in one part of the food chain have been so publicly exposed, the question many consumers may now have on food composition and labelling is "has this got horse in it?." Certainly, substitution remains a concern for local enforcement. Cheap alternatives, be that to types of fish, meat, spirits, oil or basmati rice, can be lucrative, and fraudulent or underhand introduction of them will remain under the enforcement spotlight.

The "Elliot Review" and subsequent report (2014, Interim Report December 2013), which followed the horse meat incident of 2012, has pointed to the need for a system-based approach to tackling food fraud and suggests a key role for the FSA in leading responses to "food crime." The review highlighted the need for a more coordinated and proactive approach involving central government, local authorities and the Food Standards Agency. Whether this means more resources being earmarked for food sampling and checks by local enforcers' coordinated nationally by the FSA, for central government remains to be seen, but we can expect to see food labelling, descriptions and claims in the spotlight.

3.8 Sources of further information and advice

All UK legislation, including the Food Safety Act 1990, The Food Labelling Regulations 1996, SI 1996/1499 and others referred to can be found at: http://www.legislation.gov.uk/

Advertising Standards Authority website: http://www.asa.org.uk/About-ASA.aspx

British and Irish Legal Information Institute website produces the judgement of R v Ghosh [1982] EWCA Crim 2 at: http://www.bailii.org/ew/cases/EWCA/Crim/1982/2.html

Citizens Advice consumer service is outlined at: http://www.citizensadvice.org.uk/index/getadvice/consumer_service.htm

Defra (2013, 2014) Elliot Review into the integrity and assurance of food supply networks

Dorset County Council website example of trading standards advice to business: http://www.dorsetforyou.com/tradingstandards/businesses

East of England enforcement authorities' collaboration led to the ERWIN website: http://www.everythingregulation.org.uk/Pages/home.aspx

Food Standards Agency (April 2010) *The Framework Agreement on Official Feed and Food Controls by Local Authorities.* Food Standards Agency.

Food Standards Agency (2014) *The Food Law Code of Practice (England).* Food Standards Agency.

Food Standards Agency labelling guidance may be found at their website at: www.food.gov.uk/about-us/publications/labpublications/

Food Standards Agency website pages with search facility for local enforcement: http://www.food.gov.uk/enforcement/yourarea/#.UjSFffZwZYc

FSA guidance on the Food Information Regulations 2013: https://www.gov.uk/government/uploads/system/uploads/attachment_data/file/82663/consult-fic-guidance-20121116.pdf

Government website links detailing food labelling provisions and quantity marking: https://www.gov.uk/food-labelling-and-packaging

https://www.gov.uk/food-information-to-consumers

Medicines and Healthcare Products Regulatory Agency website: http://www.mhra.gov.uk/Howweregulate/Medicines/Doesmyproductneedalicence/Borderlineproducts/

Primary Authority Scheme on the Better Regulation Delivery Office website: www.bis.gov.uk/brdo/primary-authority

Supreme Court decided case of Torfaen County Borough Council v Douglas Willis Ltd [2013] UKSC39: www.supremecourt.gov.uk/decided-cases/index.html

The Regulators' Code published by BRDO 2013 can be found at: http://www.bis.gov.uk/assets/brdo/docs/publications-2013/13-1016-regulators-code.pdf

The statutory principles of good regulation can be viewed in Part 2 (21) on page 12: http://www.legislation.gov.uk/ukpga/2006/51/pdfs/ukpga_20060051_en.pdf

Trading Standards Institute website advice material on food: http://www.tradingstandards.gov.uk/advice/advice-business-food.cfm

Website of the Trading Standards Partnership South West: http://swercots-partners.org.uk/

Enforcement of food and beverage labelling legislation: enforcement policies and codes in the UK

4

I.N. Hancock
Trading Standards Service, Dorset County Council, Dorchester, UK

4.1 Introduction

As noted in the previous chapter, a food business considering labelling requirements for UK food and beverage products, or indeed anyone researching legal requirements and enforcement in this field, may be quite understandably intimidated by the jigsaw of arrangements that exists in terms of local enforcement and how that fits into a national framework to maintain food standards.

The previous chapter described the framework of local government regulation and the national bodies responsible for coordination of enforcement at the local level, as well as summarising the relevant legislation. This chapter concentrates on enforcement policy and, in particular, the way in which those responsible for food labelling enforcement go about that task. Enforcement bodies themselves need to ensure they comply with national expectations to ensure this food enforcement jigsaw provides a more consistent approach which is supportive of businesses. The chapter also describes the consequences for businesses of failure to achieve compliance, notably if that is by repeated failure or avoidance.

Together with the previous chapter, it is hoped this chapter provides some detail around the reality of local enforcement in the context of the national framework and reassurance as to the support and advice that a business can get from a local enforcement service, whether through trading standards or environmental health professionals.

4.2 Enforcement policies relating to food and beverage labelling

Local authority enforcement is itself subject to controlling and guiding measures which seek to reduce inconsistency and provide an overall strategy to deal with

Advances in Food and Beverage Labelling. http://dx.doi.org/10.1533/9781782420934.1.51

enforcement decision making. As a result, any local authority enforcing legislation covering food and beverage labelling would be expected to have in place an enforcement policy which is available to business and residents. It would be reasonable to expect that any business having contact with a local trading standards or environmental health service could readily find a copy of that policy on the local authority's website, and numerous examples following similar formats are in evidence through an Internet search.

An authority's enforcement policy sets the tone for regulatory enforcement. They do not vary greatly, as national legislation and guidance has sought to bring approaches in line and, while it is evident that formal action levels across the country have some variance, these policies bring some transparency to the decision making process. There is broad recognition that those effectively managed compliant businesses should be supported by local authority enforcers, and that adequate protection needs to be given to compliant businesses and consumers from those not taking their responsibilities seriously.

Local authority enforcement policies, since April 2014, must "have regard to" the requirements of the revised Regulators' Code 2014. This replaced the previous Enforcement Concordat and Regulators' Compliance Code 2008. Authorities taking formal enforcement action, such as prosecution, also need to have regard to the Code for Crown Prosecutors 2013.

The earlier Regulators Compliance Code was published by the Government's Better Regulation Executive following the Hampton Review "Reducing Administrative Burdens: Effective Inspection and Enforcement." This provided a vision of the regulatory framework focused on risk and proportionality. A subsequent key review took place into the effectiveness of formal actions and sanctions against businesses, the Macrory Review "Regulatory Justice: Making Sanctions Effective." The combined effect of these reviews was to further seek to ensure that formal enforcement action was only taken in appropriate circumstances, and sought to ensure that alternative measures were explored by local enforcers to secure compliance.

The Code urged regulators to allow and encourage economic progress and only intervene where there was a clear case for protection of consumers and compliant businesses. Effort should be concentrated where the risks are greatest, and those persistently breaching requirements should feel the sanctions of the law.

4.2.1 A Revised Regulators' Code 2014

The foreword to the new Code outlines the reasoning behind introducing the changes:

> *This Government is committed to reducing regulatory burdens and supporting compliant business growth through the development of an open and constructive relationship between regulators and those they regulate. The Regulators' Code provides a flexible, principles based framework for regulatory delivery that supports and*

> *enables regulators to design their service and enforcement policies in a manner that best suits the needs of businesses and other regulated entities.*
>
> *Department for Business Innovation & Skills (2013) Regulators' Code*

The Regulators' Code 2014 outlines how regulators should address six key strands, and the Better Regulation Delivery Office (BRDO) has provided some guidance setting out how regulators address these additional burdens with the inevitable consequence of increased bureaucracy to the enforcement process including consultations, satisfaction surveys, monitoring processes and training to support these new requirements. Much of what is expected of regulators is already in place in local authority trading standards and environmental health enforcement services but this seeks to ensure the principles are consistently applied and across a wide range of what the government have termed "non-economic" regulators. There is likely to be some increased focus on the enforcement process as a result.

The approach is built on the principles found in Legislative and Regulatory Reform Act 2006, Section 21 on the exercise of regulatory functions:

Those principles are that –

> *(a) regulatory activities should be carried out in a way which is transparent, accountable, proportionate and consistent;*
> *(b) regulatory activities should be targeted only at cases in which action is needed.*

The six key strands are:

1. Regulators should carry out their activities in a way that supports those they regulate to comply and grow.
2. Regulators should provide simple and straightforward ways to engage with those they regulate and hear their views.
3. Regulators should base their regulatory activities on risk.
4. Regulators should share information about compliance and risk.
5. Regulators should ensure clear information, guidance and advice is available to help those they regulate meet their responsibilities to comply.
6. Regulators should ensure that their approach to their regulatory activities is transparent.

In terms of food businesses engaging with local authority enforcers, there may be more attempts to solicit feedback through satisfaction surveys or consultation on the regulators' approach, but the main approaches the Code promotes to are already commonplace, as this chapter hopefully illustrates.

4.2.2 Code for crown prosecutors

A further key consideration to be included in enforcement policies and to take account of in the formal enforcement process is that given by the Code for Crown Prosecutors (2013). This is issued by the Director of Public Prosecutions under the Prosecution Offences Act 1985, essentially for the Crown Prosecution Service but other prosecutors have to have regard to it. It gives guidance on general principles for bringing a prosecution to court. Prosecutors are asked to ensure that the factors leading to prosecution outweigh those that do not, as regards seriousness, culpability and circumstances such as harm caused and impact on the community.

It sets out the two stages that must be met, the evidential test and the public interest test, before a prosecution should be brought to court, underlining that there need not be a prosecution simply because offences have been committed and detected. It encourages decisions to be made at an early stage so that detailed investigation and gathering of evidence does not proceed where it is clear that a prosecution would not be in the public interest given the cost and disruption caused. In food matters, such considerations are routine and prompt alternative remedial action are sought. Throughout any investigation and court process, the Code underlines the importance of regulators continuing to review cases in the light of new evidence or changed circumstances of the parties involved, with continued consideration given to disposal of matters in an alternative way where that is appropriate. This may include changes to business structure or key personnel, for instance, or where new evidence comes to light about offences or defences.

If prosecution remains the chosen path, prosecutors are required to ensure that the offences selected, for which an information is laid and a summons issued in the case of food matters, reflect the seriousness of the matter, give adequate powers to the court to sentence, and enable the case to be presented in a clear way. Prosecutors should aim to keep it simple and not overburden the information and summonses which set out the offences. It is unhelpful for all concerned if more than ten offences are being alleged in a prosecution, even in more complex or serious cases, and for a food matter one would expect less than that. There are always other means of having matters "taken into consideration," using sample offences or by bringing relevant background evidence to the court's attention.

4.3 Local approaches to enforcement

The image portrayed in the media of trading standards enforcement activity combined with political rhetoric does little to provide a genuine picture of how trading standards authorities and their officers work. More detailed scrutiny of the role and approach of trading standards authorities reveals much more than common perceptions offer.

Assisting businesses with advice and guidance is one of the key objectives for trading standards authorities, with a variety of approaches being employed to do that. Any review of a local trading standards authority's service plans will reveal this as a vital part of the role. Often more attractive to the media, and presumably the public digesting the material, are the stories of legal sanctions being taken in the courts and of businesses, and sometimes enforcers, behaving badly. This will commonly cover seizure of goods, prosecutions, the outrage, the inadequate sanctions and the exploration into how the public have been let down by business, government or local government. The public do not want to hear about how a trading standards officer's advice has led to a smooth solution to complaints or compliance issues raised, and nor does the media or the public have the appetite for reviews of local authority initiatives put in place to provide high quality legal advice to businesses on food legislation law or other trading law issues.

Another commonly portrayed perspective is that of the jackbooted official armed with a clipboard, wasting valuable time and stifling business ideas and activity. In recent years, there has been considerable political appeal in talk of banishing red tape and ending the tick box approach to inspections. These historic stereotypes fail to recognise the shift away from routine visits by trading standards officers to all but high risk food business, and that those visits aim to help ensure compliance through advice and guidance in all but the most extreme cases. Trading standards services risk-rate their businesses such that focus for any proactive routine visits is on those businesses which have the most impact on the public, and the standards and composition of the food supplied. This means that significant focus is on those manufacturing and processing, importers (particularly those importing goods from outside the European Community), and those activities that intelligence systems point to as areas demanding intervention. The use of intelligence to direct enforcement activity continues to increase and hence allow trading standards authorities to channel increasingly limited resources towards those business practices causing greatest concern for consumers and compliant businesses.

For the vast majority of UK businesses their experience of trading standards officers is likely to be the discussion of how particular legislation affects them and guidance on how to meet those customer expectations that have been encased in UK law, and that usually means European Community law when considering food and beverage labelling. This may be following a request from a business, whether a problem has been raised by a consumer, another trader, another trading standards service, or other enforcement body in the UK or from elsewhere in Europe. Alternatively, this may be from a visit, often pre-arranged at a time more convenient to the business, or perhaps as part of a targeted survey into areas of compliance that intelligence point to as being one for concern.

Advice and guidance to business is the preferred route to finding solutions to problems of non-compliance. Local authorities have statutory duties to enforce food standards, labelling and compositional requirements, but the vast majority of officers' time is spent securing legal and workable outcomes to regulatory problems. And it is not of course all about seeking "compliance" in its own right. Through compliance with trading law, especially in the area of food and drink labelling, products are saleable through major suppliers and local producers, and processors are able to access national and international markets. Optional labelling requirements may bring added consumer appeal to products, whether that is due to giving prominence to particular sources and origins, or correctly applied nutritional emphasis that can increase sales. Compliance should mean the consumer, the business, business generally, and the enforcement authority will all be satisfied.

Providing best advice and guidance to businesses based in a local authority's area is a key activity of trading standards services and there are various mechanisms in place to enhance the simple visit or response to business request for information. The needs clearly vary, from very well informed large and experienced companies to small enterprises that find themselves faced with hurdles to the smooth growth of their business. The key routes to assistance from a local trading standards authority are outlined below.

4.3.1 Web-based information

Many local services buy into nationally available web-based advice products such as the Trading Standards Institute's "Trading Standards Broadcast" product. This allows material explaining food labelling and trading law to businesses to be regularly reviewed and updated by trading standards professionals. These appear as trader guidance through participating local authorities' own websites and through the Trading Standards Institute's main website:

http://www.dorsetforyou.com/tradingstandards/businesses
http://www.tradingstandards.gov.uk/advice/advice-business-food.cfm

There is also free access to web-based material that gives key advice based on business sector and type – see http://www.everythingregulation.org.uk/Pages/home.aspx. This excellent site is a result of a collaborative project by East of England local enforcement authorities started in 2009. Again, this can often be accessed through local authorities' own websites as a gateway. This serves to focus on the key areas of law that a business needs to pay attention to, and gives an indication of the priority given to certain legislation. This can be enormously helpful to new small food businesses that are unfamiliar with the range of regulatory requirements, and allows priority areas for compliance to be seen easily. Further labelling guidance is available from the FSA website.

4.3.2 Advice to businesses

Most trading standards authorities still provide advice to business based in their local area at no charge. The service is part of the activity that seeks to ensure compliance with the laws that the local authorities have a statutory duty to enforce, and for which local authorities are funded. As budgetary cuts continue to put pressure on services to reduce activity, or find any form of cost recovery or income, there are likely to be increasing examples of trading standards authorities charging for advice in some form.

Advice is available via bespoke telephone helplines advertised, by emailing services, and through officers arranging routine visits or being requested to visit by a business due to particular problems arising. Many unfamiliar with food law regulators such as trading standards fear contacting "the authorities." However, contact is encouraged by trading standards services. A business seeking assistance is likely to be met with a very positive approach seeking to avoid formal enforcement action, especially where they are responding to a trader's proactive approach. Business needs reassurance that the stance taken by authorities is to try and help remove hurdles and smooth over concerns by effective advice on food law requirements. It is virtually unheard of for a business to face formal enforcement action where they have sought assistance with a food labelling or composition issue. Of course, there will be instances where authorities cannot turn a blind eye to certain breaches, and some breaches will attract the need to recall goods, but in many cases breaches can be dealt with by agreed action being taken over a given time period.

4.3.3 The Home Authority Principle

The Home Authority Principle is a well understood and engrained approach which has been embraced by UK trading standards and food authorities for over 25 years. It is also supported by environmental health authorities enforcing food labelling law, and is a principle now supported and administered jointly by the professional bodies for trading standards officers (the Trading Standards Institute) and environmental health officers (the Chartered Institute of Environmental Health). Under the Home Authority Principle (HAP), local services can focus their attention and resource on goods and services emanating from their local authority area. Most importantly, it allows multi-site businesses or businesses selling goods or services nationally or trading across several local authority areas to engage with UK authorities, on food and beverage labelling or other trading law issues, and to receive consistent advice from a trusted source.

This enables both businesses and local authority services to reduce duplication of efforts in explaining and investigating trading law issues. It has proved particularly helpful in the area of food and beverage labelling, where the detail of regulatory requirements mean that advice, guidance and improvement over agreed timescales is so often the enforcement outcome sought. Formal enforcement action on food composition and labelling remains very rare, and this is because officers can see the evidence that Home Authority advice usually brings compliance through agreement. With pressure on staff resource, it is also an attractive option to refer problems found at a local level to the Home Authority service for action, in the knowledge that the responsibility for achieving an appropriate resolution is also discharged.

Where offences are found locally through complaints, sampling, inspection or other investigation, the option still remains for that authority to consider and take formal enforcement action where that is appropriate, such as with more serious breaches or where it is known that previous Home Authority advice has been ignored. In such cases an enforcing authority may make direct contact in the form of a formal interview under caution, in person or by correspondence. Any authority's investigation would be expected to include contact with a business's Home Authority at an early stage when a problem is found, and they will take account of any advice previously given.

The Home Authority, as the main link with a business on behalf of other local authorities, will maintain a record of relevant incidents, company policies, diligence systems and advice given. Those involved in seeking to resolve issues by negotiation and agreement have to recognise that an investigating authority cannot be prevented from taking its own legal action if it still believes that is appropriate.

So where is the Home Authority, and how do officers know which authority, and often which individual officer is acting as this key contact point for a business? An agreement to act as Home Authority is usually based on where the head office is in the UK, or certainly where those making the relevant key decisions that affect compliance are based. The liaison point needs to be agreeable to the business and the local authority concerned. A business may have one or more production points in different areas. While those production points may be subject to physical checks at the place of production or import by what is referred to as the "originating authority," changes to labelling, composition, processes or policy would be discussed and agreed via the Home Authority. Formal Home Authority arrangements are underpinned by signed agreements between local authorities and businesses based in their geographical area but in any event the approach is widely followed. Local authority services register the Home Authority arrangements they have centrally.

The HAP developed mainly under years of enthusiastic stewardship and promotion by the Local Authority Co-ordinating Body on Food and Trading Standards (LACOTS), funded by the Local Government Association. LACOTS transformed and later withdrawal of funding led to the Home Authority database being hosted now with the Trading Standards Institute (TSI) via their secure TS Interlink on-line product for local authorities. The existing system was enabled by an agreement between TSI, the Chartered Institute of Environmental Health and the BRDO, the last being behind the newer statutory scheme which has been aiming to achieve similar aims.

The HAP is a longstanding, well understood and well used voluntary approach used by local food enforcement services to bring a high level of public protection and product compliance at minimum cost to business and local authorities. There is, however, a newer statutory option to consider which has received significant financial support and promotion as a key plank to the work of the BRDO: the Primary Authority Scheme (PAS).

4.3.4 The Primary Authority Scheme

The PAS has taken the HAP and given it a statutory footing that enforcing authorities must have regard to. It is an option to consider for any food business trading beyond a single local authority area. The take up of this scheme has been variable despite significant central government funding, supporting BRDO, being applied to promote the scheme and encourage local authorities and businesses to engage with it.

In short, the PAS provides the statutory certainty that some businesses, particularly companies such as the large supermarket chains, were looking to address their concerns about regulatory burden. An enforcement officer now has to have regard to advice given under the PAS and to any inspection plans agreed, thus restricting additional enforcement activity except in particular circumstances. A key difference is also that businesses may pay the local authority for its advice and guidance under an agreed contractual arrangement which brings certainty to the resource made available to support business's needs. With constraints on local authority service budgets, although payment received is supposed to be on a cost recovery basis, this can be an attractive option to support this work and provide a more resilient workforce structure.

The PAS exists to give a business a statutory partnership with one local authority at a convenient location and with enforcement staff resource being committed by the local authority, which can be paid for by the business to support that partnership. In a similar way to the HAP, the PAS seeks to ensure a business is provided with robust, reliable advice which other authorities must take account of when addressing non-compliances they find or are brought to their attention. The arrangements can cover a range of trading standards or environmental health law, or be just on particular areas where it is perceived to be most beneficial, indeed food composition and labelling for example. It is for the business to decide the level of support needed and it can then negotiate a contract with the local authority concerned to put certain commitments in place to suit the business needs.

A key aspect of this arrangement is that it enables a business to agree on an inspection plan with an enforcement authority. This plan is then registered on the secure web-based Primary Authority IT System, which is also where any PAS arrangements are recorded. This allows for a reasonable level of inspection to be agreed, and limit unnecessary further disruption by inspection at the businesses outlets and hopefully avoided repeated local checks on the same compliance issues. This will be of more use to multi-outlet retailers than, say, a food manufacturer or importer.

By April 2014 there were 1500 businesses covered by the PAS, with 120 local authorities involved. With over 200 Trading Standards authorities in the UK and over 400 local authorities enforcing legislation covered by the scheme, it is clearly not a universally popular option. BRDO have been continuing to explore the reasons for this, and to extend and further promote the scheme. The scheme has been extended to trade associations and franchisees through single contact points. More can be found on the PAS on the BRDO website: www.bis.gov.uk/brdo/primary-authority.

4.4 The implications of a failure to meet legal obligations relating to food and beverage labelling

The commitment of local authorities to resolving food and beverage labelling issues by way of advice and improvement is clear. It is the usual form of intervention if a breach is identified but, if an acceptable resolution cannot be arranged with a business within a suitable timescale, more formal action may be considered. It is that possibility of formal action and potential reputational damage that can flow from it that often has the effect of achieving a required change. So what are the formal action options for food and beverage compositional and labelling offences?

Following intervention by an enforcing authority, some investigation may follow unless it is simply referred to the Home Authority or Primary Authority. An investigation may be concluded with a simple letter of warning outlining the problems found. Some authorities seek acknowledgement of such warning letters or other communications to ensure the message is received. A warning may be sent out quite quickly after a breach has been identified, but the investigation may not rule out formal action until questions are asked "under caution" in line with the Police and Criminal Evidence Act Codes of Practice. For those unfamiliar with a trading standards or other regulatory investigation, this can come as something of an unpleasant surprise. However, it has become well established that where offences have been committed and discussions take place between enforcement officers and individuals or representatives of companies suspected of what are criminal offences, these discussions are unfair on potential defendants.

Those suspected of committing offences are afforded the protection of being cautioned that:

> *You do not have to say anything. However, it may harm your defence if you do not mention when questioned something which you later rely on in court. Anything you do say may be given in evidence.*

Once an offence is suspected any comments not made under caution may be excluded in trial and hence investigations may appear to suddenly become very formal. Officers will seek to ask questions in a face-to-face interview under caution that is likely to be recorded on disc or tape, with the recordings being carefully handled to ensure their integrity is preserved.

On occasion, authorities may put questions to alleged offenders in writing, under caution, as a practical alternative but this is often not appropriate and can lead to further correspondence.

Having obtained answers to questions under caution, the local authority will consider what action to take. The conclusion may be that no offence has been committed by the parties suspected, or that a statutory defence has been established. It may be that evidence of an offence is clear and no full defence is apparent, but that the explanation of the circumstances is such that a prompt written warning is still the outcome. There is no requirement to prosecute unless as well as the evidence being clear the public interest would be served by taking a matter further. Local authorities will usually use their publicly available and agreed enforcement policies to aid such decisions. After consideration of the circumstances by senior service managers, legal services teams, or even in some cases by panels of elected Councillors of the local authority, the decision will be to deal with cases by written warning, a Simple Caution, or prosecution.

4.4.1 Simple Cautions

A Simple Caution is a possible outcome where there has been an admission of offences and there is sufficient evidence to prosecute but the circumstances are such that it is an appropriate solution which avoids formal court action and the associated costs for both defendant and prosecutor. Simple Cautions are kept on record by local authorities and can be referred to if court action results are required in future, and they can be recorded on a national database maintained on behalf of the National Trading Standards Board (Sanctions Information Database – SiD). There have been occasions where Simple Cautions have been signed and an offer made to pay costs incurred by the prosecution in the investigation as part of that solution. While many authorities may find this slightly unsavoury, some local authority services can see this as a legitimate attempt to cover costs while also coming to a satisfactory conclusion. With increasing pressures on local authority budgets, perhaps this may become a more common arrangement for the offender to pay without a formal prosecution.

4.4.2 Prosecution

What are the consequences of prosecution? There can be little doubt that for food labelling and composition law contraventions the fact of a conviction against a company, or a business, is in itself very negative in terms of reputation. Depending on the issue, the ensuing negative publicity and damage will be a significant aspect to any case. Except in the most fraudulent cases, the penalty in such a prosecution is extremely likely to be a fine and experiences are that it will range from several hundred to a few thousand pounds in each case. The level of fine can escalate in some cases involving larger businesses and where there is evidence of repeated failures and deliberate acts.

Depending on the business structure, a prosecution will usually be against a limited company or the proprietor or any number of proprietors of an unincorporated business. However, individual directors may be prosecuted where they are particularly responsible and the legislation allows it. Prosecutions of individual directors will depend on particular roles and levels of involvement in what is the focus of the prosecution. Also individual employees may be prosecuted if their fault has caused a particular problem. Except in the most serious cases, prosecutions against individual employees who are not part of the controlling mind of the business are very rare in this field of law.

Although the general experience is of relatively low level fines, there are maximum penalties to be aware of which can include imprisonment. Fines for food standards and labelling matters can be subject to maximums of £5000, or £20,000 in relation to some offences, with imprisonment also being a rarely used option in food cases. Cases under the Food Safety Act 1990 or the Food Labelling Regulations 1996 can be dealt with in a magistrates' court or, in more serious cases, sent to a crown court where maximum fines can be unlimited and terms of imprisonment of up to 2 years considered.

It may be that if a problem arises with the labelling, description, advertising or composition of a food or beverage, especially a serious one such as the recent problems of horse meat being substituted in products, that other legislation would be used, such as the Fraud Act 2006. If breaches merit it, there is no doubt the courts have sufficient powers of sentence to deal with irresponsible or fraudulent food producers.

Alternative punishments to imprisonment of individuals can include suspended sentences or community orders to carry out unpaid work for example. An offending company or individual may also be found guilty or plead guilty but be given an "absolute discharge" if the court finds the matters are so trivial or the prosecution itself is sufficient in the circumstances. A "conditional discharge" is another option where a conviction stands, and costs may be payable, but no additional penalty results. This is conditional on no further offences being committed in a stated period, which may be up to 3 years.

Sentencing guidelines are not available to magistrates or judges on food law but do exist for certain areas of trading standards law dealing with counterfeit goods, underage sales of alcohol and animal welfare offences. Sentencing must take account of appeal cases on sentencing, but again there is little to guide on food law.

Guilty parties are also likely to be asked to pay the costs of the local authority in bringing the prosecution to court, including officer investigation and report preparation time, analytical and sampling costs, solicitor time and other legal fees. In the case of an early guilty plea, these will be in the region of £1000–3000. However, a contested case with a trial may rise to £5000–10,000, even in the magistrates' courts, and beyond that if appeals to a higher court are involved. Occasionally, large and complex cases taken to trial can lead to prosecution costs in excess of £50,000 but such cases are very rare. The costs including disruption to usual business involved in dealing with a prosecution are likely to be as much of a consideration.

It is not just the business owner, the partners, or a limited company that can face prosecution. There is the personal liability of individual directors, managers or the secretary of a body corporate to bear in mind. These company officers "or any person who was purporting to act in any such capacity" may be prosecuted if it can be shown that offences under the Food Safety Act 1990 are due to the "consent, connivance or attributable to the neglect" of that person (Section 36, Food Safety Act 1990). This is irrespective of whether action is taken against the company, and can be a useful provision for enforcers where a company has ceased trading.

Individual employees or indeed printers or others involved may be prosecuted (under Section 20, Food Safety Act 1990) for any offences under the Food Safety Act if it is due to "an act or default" of theirs, whether or not the company, an owner or partner is prosecuted as a principal offender. The defence of innocent publication is available (Section 22) for those "whose business it is to publish or arrange to publication of advertisements" as long as the advertisement was received in the ordinary course of business and he did not know, and had no reason to suspect, that publication would be an offence. That defence would not assist if the publisher were told the advertisement was a breach, by an enforcement officer for example.

In most food cases, including food labelling matters, the general defence provisions are as key a consideration as the offence.

Under the Food Labelling Regulations, Regulation 48, the sections dealing with corporate bodies, "act or default" and innocent publication are also brought into force and may assist or indeed serve to further focus responsibility appropriately.

4.4.3 Other business costs

Any matter where independent or in-house legal advice is sought by businesses or individuals involved will of course run up its own costs, and experience is that contested matters that go to trial usually lead to defence legal costs that outweigh those of the prosecution considerably. If a defence against an enforcement authority is successful then costs are usually awarded from central court funds unless the prosecution has been handled badly, in which case costs can be ordered against the prosecuting body. The court need not make an order for costs to be paid out of central funds if, for example, the defendants have conducted themselves in such a way that they could have avoided the case being brought before a court.

Aside from any court action, any failures to deal with the supply of food that is inadequately labelled, misdescribed, or composed of something it should not be may lead to the costs of recalling product, either voluntarily or as the result of the FSA's food alert process. When problems arise and are identified at an early stage, trading standards authorities, or environmental health officers, can be helpful in agreeing remedial action over a reasonable time period, so avoiding the wasteful process of unnecessarily withdrawing stock and discarding packaging for example.

4.5 Future trends

As noted earlier, the changing consumer protection landscape will continue to see improved coordination of local consumer protection enforcement generally at national and regional level, and the opportunity is there for this to embrace food work to a greater extent. This is against a background of public sector cuts that will see further reductions to local food authorities, reducing their ability both to advise and support business and indeed take formal action. The Regulators' Code introduced in April 2014 gives further reason to focus the reduced enforcement resources on those high risk activities and businesses.

The jigsaw of food labelling enforcement will continue to change shape as local arrangements to merge services continue, based on budget saving opportunities and local politics rather than any attention to effective local food enforcement. Finances may also see authorities seeking to recover costs for business advice, whether that is through the PAS or other local charging arrangements.

The recommendations from the Elliot Report on the review that followed the horse meat incident are likely to give increased scrutiny to food fraud or any "food crime" the report refers to. A system-based approach to tackling food fraud has been called for that gives food safety and food crime prevention absolute priority over other objectives. Reference to a zero tolerance approach suggests that misleading labelling and false claims will be challenged and there will be an expectation and maybe increased commitment from enforcement authorities to ensure the confidence in the food chain is not undermined. The review suggests a significant change in culture and coordination is required to address the threats of fraudulent activity in the food sector. The Food Standards Agency, supported on the ground by local authority enforcement, is likely to lead that.

Reputable businesses will continue to get the support and advice that has developed, but once problems are unearthed or true criminal behaviour is suspected, it is likely a harder line will be demanded, involving a more thorough and coordinated response from a range of enforcement authorities.

4.6 Sources of further information and advice

Dorset County Council website example of trading standards advice to business: http://www.dorsetforyou.com/tradingstandards/businesses

Defra (2013, 2014) Elliot Review into the integrity and assurance of food supply networks.

East of England enforcement authorities' collaboration led to the ERWIN website: http://www.everythingregulation.org.uk/Pages/home.aspx

Primary Authority Scheme on the Better Regulation Delivery Office website: www.bis.gov.uk/brdo/primary-authority

The Regulators' Code published by BRDO 2013 can be found at: http://www.bis.gov.uk/assets/brdo/docs/publications-2013/13-1016-regulators-code.pdf

The statutory principles of good regulation can be viewed in Part 2 (21) on page 12: http://www.legislation.gov.uk/ukpga/2006/51/pdfs/ukpga_20060051_en.pdf

Trading Standards Institute website advice material on food: http://www.tradingstandards.gov.uk/advice/advice-business-food.cfm

Part Two

Trends in labelling relating to nutrition and health

Nutrition and related labelling of foods and beverages: the case of the USA

5

L. Fitzpatrick
Ashbury Labelling, London, UK

5.1 Introduction

This chapter builds on Chapter 2, which provides a general introduction to the regulatory framework governing food and beverage labelling in the USA. As noted in that chapter, although the stated principles of protecting public health and consumer interests are the same in the USA as in the European Union, there are many contrasts. This chapter can only provide an overview of nutrition and related labelling regulation in the US, and will focus mainly on regulations developed and enforced by the Food and Drug Administration (FDA).

This chapter first looks at general requirements related to nutrition labelling, including information on types and levels of nutrients, serving sizes, and daily values. It then outlines general format requirements for labels. After discussing the particular cases of labelling related to restaurants and alcoholic beverages, the chapter discusses issues related to voluntary labelling statements or claims made by manufacturers to advertise particular aspects of a food product that may affect purchase. These include claims about nutrient content, health claims, statements about organic or genetically-modified (GM)-free foods, gluten-free claims, and claims relating to "natural" foods.

5.2 Nutrition labelling: general requirements

The Nutrition Labeling and Education Act (NLEA) of 1990 amended the Federal Food, Drug and Cosmetic (FD&C) Act to require the mandatory nutrition labelling of most foods. On 6 January 1993, the FDA issued nutrition labelling regulations to enforce these requirements. These regulations are laid down in Title 21 of the Code of Federal Regulations (21 CFR) Part 101.9. The provisions on nutrition labelling are extremely extensive and detailed, and it is only possible to provide a basic overview for the purposes of this chapter.

Advances in Food and Beverage Labelling. http://dx.doi.org/10.1533/9781782420934.2.67

In parallel to the FDA requirements, the Food Safety and Inspection Service (FSIS) requires mandatory nutrition labelling of all multi-ingredient and heat-processed meat and poultry products, which are largely equivalent to those of the FDA.

Nutrition information must be provided for all products intended for human consumption and offered for sale, unless specifically exempted. As a few examples, exemptions are provided for small businesses, raw fruits and vegetables, unit containers in a multi-unit retail food package that provides nutrition information, and foods that contain insignificant amounts of all of the nutrients and food components required to be declared. An insignificant amount is that which allows a declaration of zero in nutritional labelling. For total carbohydrate, dietary fibre, and protein it is the amount that allows a declaration of "less than 1 g." Examples of such foods include coffee beans (whole or ground), tea leaves, unsweetened coffee and tea, condiment-type dehydrated vegetables, flavour extracts, and food colours.

A nutrition claim or other nutrition information on the label, labelling, or in advertising negates any exemption that was provided. For packaged food, the required information must appear on the label. For non-packaged food, nutrition information must be given at the point of purchase (e.g. on a counter card or tag attached to the product), or the information may be given in a booklet, loose-leaf binder, or other appropriate format.

5.2.1 Nutrients

The nutrition label must include information about the levels of the following nutrients:

- Total calories
- calories from fat
- calories from saturated fat*
- total fat
- saturated fat
- trans fat
- polyunsaturated fat*
- monounsaturated fat*
- cholesterol
- sodium
- potassium*
- total carbohydrate
- dietary fibre
- soluble fibre*
- insoluble fibre*
- sugars
- sugar alcohol*
- other carbohydrate*
- protein
- corrected amount of protein*
- vitamins and minerals (vitamin A, vitamin C, calcium, and iron must be declared even when they contribute zero per cent)

*Voluntary

The nutritional values must be declared per serving by weight and as a percentage of the "Daily Value" (DV) (except for sugar and protein); an additional declaration per 100 g is optional. Vitamins and minerals are declared only as a percentage of the DV. Rounding rules for declaration of nutrients are also laid down. Where any of the nutrients meet the requirement of a declaration of zero, a declaration may be omitted, provided that a statement "Not a significant source of [named nutrients]" is placed at the bottom of the table of nutrient values in the same type size.

5.2.2 Serving size

According to Part 101.9(4)(b)(1), the term *serving size* or *serving* means an amount of food customarily consumed per eating occasion by persons 4 years of age or older, which is expressed in both common household and metric measures in brackets. Serving size shall be determined by the "Reference Amounts Customarily Consumed Per Eating Occasion," laid down for a large number of food categories in 21 CFR Part 101.12.

The intention is to provide the consumer with a relatable, visual unit of measurement for the nutrients consumed per portion. Some examples of declarations of serving sizes are:

- Cookies "1 cookie (28 g)" or "1 cookie (28 g/1 oz.)"
- Juices "8 fl. oz. (240 mL)" or "1 cup (240 mL)" for multiservings
- Soft drinks containers (e.g. "1 can") for single servings
- Grated cheese "1 tablespoon (5 g)" or "1 tablespoon (5 g/0.2 oz.)"

Methods for determining serving size are laid down and vary by food. Determination of serving size for discrete unit weights in relation to the Reference Amounts Customarily Consumed Per Eating Occasion (RACC), is as shown in Table 5.1 below.

Products in large individual units are divided for consumption including unprepared products (for example, fractions of a cake, pie, pizza, melon).

Serving size is the fractional slice of the food that most closely approximates the reference amount. When expressing a fractional slice, manufacturers must use one-half to one-sixth fractions, or smaller fractions that can be generated by further division by 2 or 3.

For non-discrete bulk products (e.g. breakfast cereal, flour, sugar, dry mixes, and concentrates), serving size is the amount in household measure that most closely approximates the reference amount.

When the serving size falls exactly halfway between two serving sizes, it must be rounded up to the next incremental size. There are further rules for variety packs and single-serving packages.

The statement of the number of "Servings Per Container" is required underneath the declaration of a serving. In general, the number of servings is rounded to the nearest whole number, except for two to five servings, which is rounded to the nearest one-half serving (and using the term "about").

Table 5.1 **Determination of serving size according to quantity in relation to Reference Amount Customarily Consumed**

Quantity of product in relation to RACC	Serving size
≤50%	Number of whole units that closely approximate the RACC
50–67%	1 or 2 units
67–200%	1 units
≥200%	1 unit if whole unit can be reasonably consumed on one single occasion
If RACC ≥ 100 g or mL and are individual units in multiserving package, if unit < 150% but <200% of RACC	1 or 2 unit
Products varying naturally in size (e.g. pickles or gherkins, shellfish)	Amount in ounces that closely approximate RACC
Two or more foods packaged for consumption together and when the main ingredient is a discrete unit (e.g. pancakes and syrup)	Units of main ingredient plus proportioned minor ingredients used to make RACC for combined product
Several individual serving containers labelled individually	1 unit
Maraschino cherries	1 cherry with metric measure of average weight of a medium-sized cherry given in brackets

Table 5.2 **Daily reference values**

Macronutrient	DRV
Fat	65 g
Saturated fatty acids	20 g
Cholesterol	300 mg
Total carbohydrate	300 g
Fibre	25 g
Sodium	2400 mg
Potassium	3500 mg
Protein	50 g

5.2.3 Daily values

FDA established two sets of values: Reference Daily Intakes (RDI) and Daily Reference Values (DRV) for use in declaring the nutrient content of a food under the heading of "Daily Value" on the label (this single term is used to designate both the DRVs and RDIs). DRVs are established for eight nutrients and food components considered to be important for the maintenance of good health, as shown below in Table 5.2, in 21 CFR 101.9(c)(9).

Table 5.3 **Reference daily intakes for essential vitamins and minerals**

Substance	RDI	Substance	RDI
Vitamin A	5000 International Units (IU)	Biotin	300 μg
Vitamin C	60 mg	Pantothenic acid	10 mg
Calcium	1000 mg	Biotin	300 μg
Iron	18 mg	Phosphorus	1000 mg
Vitamin D	400 IU (International Units)	Iodine	150 μg
Vitamin E	30 IU	Magnesium	400 mg
Vitamin K	80 μg	Zinc	15 mg
Thiamin	1.5 mg	Selenium	70 μg
Riboflavin	1.7 mg	Copper	2.0 mg
Niacin	20 mg	Manganese	2.0 mg
Vitamin B_6	2.0 mg	Chromium	120 μg
Folate	400 μg	Molybdenum	75 μg
Vitamin B_{12}	6 μg	Chloride	3400 mg

RDIs are provided for vitamins and minerals as shown below in Table 5.3 in 21 CFR 101.9(c)(8)(iv).

5.3 Formats for nutrition labels

Various formats for the Nutrition Facts table are presented that may be used depending upon the product, but the following illustrates general requirements. The illustration below in Figure 5.1 (Nutrition Label Format) indicates the suggested typesetting specifications for a "Nutrition Facts" (Standard) label to be considered conspicuous and adequately formatted.

Format requirements are specified in 21 CFR 101.9(d) as follows:

1. *Overall formatting*: Nutrition Facts panel is boxed with all black or one color type printed on a white or neutral background.
2. *Typeface and size*: Use six point or larger Helvetica Black and/or Helvetica Regular type. In order to fit some formats the typography may be kerned as much as –four.
   Key nutrients & their % DV are set in eight-point Helvetica Black (but "%" is set in Helvetica Regular).
   "Nutrition Facts" is set in either Franklin Gothic Heavy or Helvetica Black to fit the width of the label flush left and flush right.
   "Serving Size" and "Servings per container" are set in eight-point Helvetica Regular with one point of leading.
   The table headings (for example, "Amount per Serving") are set in 6 point Helvetica Black.
   Absolute measures of nutrient content ("1 g") and nutrient subgroups are set in eight-point Helvetica Regular with four points of leading.

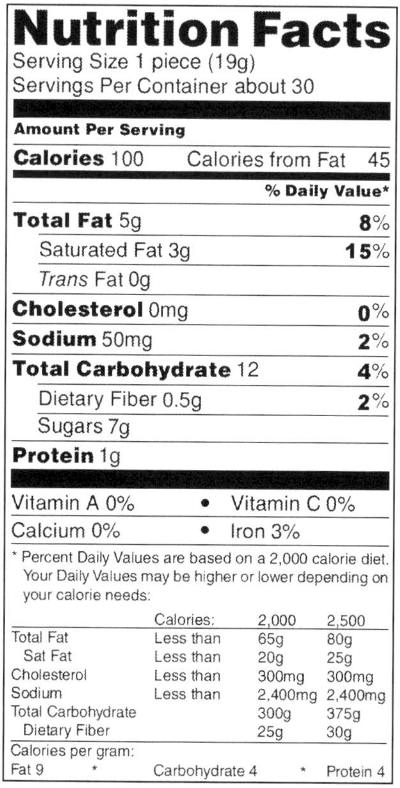

Figure 5.1 An example of a "standard" Nutrition Facts panel format with the full footnote.

Vitamins and minerals are set in eight-point Helvetica Regular, with four points of leading, separated by ten point bullets.

All type that appears under vitamins and minerals is set in six point Helvetica Regular with one point of leading.

3. *Rules*: A seven-point rule separates large groupings. A three point rule separates calorie information from the nutrient information.

   A hairline rule or one-fourth point rule separates individual nutrients, as shown in the example. The top half of the label (nutrient information) has two points of leading between the type and the rules, the bottom half of the label (footnotes) has one point of leading between the type and the rules.
4. *Box*: All labels are enclosed by ½ point box rule within three points of text measure.

All of the above are not mandatory, but FDA urged use of them in the interest of uniformity. Although the preference is for all black or one colour type printed on a white or neutral background, it is also possible to invert the typeface (the table may be printed in white or contrasting colour against a dark background if this is in keeping with the colour scheme of the rest of the label). Different fonts are rarely seen, but may be used provided they are legible. The "Nutrition Facts" heading must be the largest type size in the nutrition label (i.e., it must be larger than eight point, but does not have to be 13 point).

There are various exemptions from the standard format, based upon nutrient content (the "Simplified" format), and package format ("aggregate" format for variety packs) and size. Two examples are given below.

For foods in packages that have a total surface area for labelling of 40 or less square inches (258 sq. cm), the label may be modified as follows:

- **a.** by presenting the information in tabular/linear format rather than in a vertical format if the product has total surface area for labeling of less than 12 square inches (77 sq. cm) or if the product has a total surface area for labeling of 40 or less square inches (258 sq. cm) and the package shape or size cannot accommodate a column display on any panel. Table format must, whenever possible, be used in preference to a linear display. In that case, sub-components must be listed in parentheses after principal components (e.g. saturated fat must be declared in parentheses after total fat).
- **b.** by using certain abbreviations, for example "Serv size" for "Serving size," "Fat cal" for "Calories from fat," "Sat fat" for "Saturated fat."
- **c.** by omitting the footnote and caloric conversions and declaring the statement "Percent Daily Values are based on a 2000 calorie diet" at the bottom of the label; if the term "Daily Value" has not been spelled out, the statement that "DV" represents "Daily Value" is required.
- **d.** by presenting the nutrition information on any panel.

The tabular format for small packages is illustrated in Figure 5.2. A linear format for small packages using abbreviations is illustrated in Figure 5.3.

Nutrition Facts	Amount/Serving	% DV*	Amount/Serving	% DV*
Serv. Size 2 tbsp (28g)	**Total Fat** 9g	14%	**Total Carb.** 1g	0%
Servings about 10	Sat. Fat 2.5g	13%	Fiber 0	0%
Calories 90	*Trans* Fat 0g		Sugars 0g	
Fat Cal. 80	**Cholest.** 15mg	5%	**Protein** 1g	
	Sodium 120mg	5%		
*Percent Daily Values (DV) are based on a 2,000 calorie diet.	Vitamin A 20% • Vitamin C 2% • Calcium 4% • Iron 0%			

Figure 5.2 An example of a Horizontal Nutrition Facts panel for small packages, adapted from a small container of dip with the lid used as Information Panel.

Nutrition Facts Serv. Size: 1 Pack (14g) Serv. Per Container: 1 Amount Per Serving: **Calories** 80, Fat Cal. 60, **Total Fat** 7g (11% DV), Sat. Fat 1.5g (7% DV), Trans Fat 0g, **Cholest.** 0mg (0% DV), **Sodium** 30mg (1% DV), **Total Carb** 3g (1% DV), Fiber 1g (4% DV), Sugars 1g, **Protein** 3g, Vitamin A (0% DV), Vitamin C (0% DV), Calcium (0% DV), Iron (0% DV), Percent Daily Values (DV) are based on a 2,000 calorie diet.

Figure 5.3 A linear format Nutrition Facts panel for small package, with abbreviations. Label is a single-serving sachet of peanut spread.

5.4 Nutrition labelling in restaurants and for alcoholic beverages

Since there are different issues compared to those for standard food packaging, there are particular requirements for labelling in the case of restaurants and alcoholic beverages.

5.4.1 Nutrition labelling in restaurants

Although NLEA contained a specific exemption for foods served in restaurants or other catering establishments, certain states have had regulations in place mandating provision of Nutrition Facts in restaurants (for example, California). However, Section 4205 of the Patient Protection and Affordable Care Act of 2010 ("Affordable Care Act"), signed into law on 23 March 2010 establishes Federal requirements for nutrition labelling of standard menu items for chain restaurants and retail food establishments with 20 or more locations doing business under the same name, and chain vending machine operators. Some provisions depend on FDA to issue rules before they become mandatory.

It is already mandatory for chain retail food establishments to:

- Disclose the number of calories in each standard menu item on menus and menu boards,
- Provide written nutrition information to consumers upon request,
- Provide a "prominent, clear, and conspicuous" statement that written nutrition information is available, and
- Provide calorie information (per serving or per food item) for self-service items and food on display, on a sign adjacent to each food item.

Further rules, including those on putting calorie information in the context of a total daily caloric intake and disclosing nutrient information for items listed as single menu items but available in different flavours or combinations, have yet to be issued.

5.4.2 Nutrition labelling for alcoholic beverages

Most alcoholic drinks are also outside the scope of NLEA, being under TTB's (Alcohol and Tobacco Tax and Trade Bureau) authority instead of FDA's. TTB already had a policy in place under Ruling 2004–1, Caloric and Carbohydrate Representations in the Labelling and Advertising of Wine, Distilled Spirits, and Malt Beverages (7 April 2004). TTB proposed amending parts of the TTB regulations to set requirements for mandatory alcohol content labelling for products for which this is not currently required, and for the presentation of certain calorie and nutrient information in a mandatory Serving Facts panel for all alcohol beverage products. Under the proposal, a Serving Facts statement would include information about the number of calories, as well as the number of grams of carbohydrates, fat, and protein per serving, in a box entitled "Serving Facts." TTB has issued interim guidance on the use of "Serving Facts" statement on labels and in advertisements for alcoholic beverages. Pending the completion of the rulemaking, it has issued this document to

Serving Facts	
Serving Size	5 fl oz (148 ml)
Servings Per Container	5
	Amount per serving
Alcohol by volume	14%
fl oz of alcohol	0.7
Calories	120
Carbohydrate	3g
Fat	0g
Protein	0g

Figure 5.4 A TTB "Serving Facts" label for a wine bottle including alcohol content.

provide guidance on the voluntary declaration of Serving Facts. The guidance gives examples of various formats which may be used for different beverages.

Figure 5.4, above illustrates a Serving Facts for a 750 mL bottle of wine with 14% alcohol by volume, with optional declarations of percent by volume alcohol content and alcohol in fluid ounces.

It should also be noted that FDA has issued four Advance Notices of Proposed Rulemaking, proposing to amend and update elements of NLEA regarding the prominence of information on calories, revision of serving sizes and RACCs, and revisions of DVs/RDIs. FDA is also studying Front-of-Pack nutrition information, as this is not currently specifically regulated.

5.5 Voluntary labelling statements: nutrient content claims

21 CFR Part 101.13 lays down general requirements for nutrient content claims. These are claims that expressly or implicitly characterize the level of a nutrient that has a DV established by the FDA for nutrition labelling. General rules on making claims and permitted claims are laid down in the regulations.

An expressed claim is a direct statement about the level or range of a nutrient in a food. An implied claim suggests that a nutrient is absent/present in a certain amount (e.g. "high in oat bran") or suggests that the food, because of its nutrient content, may be useful for dietary health and is made in conjunction with an explicit claim or statement (e.g. "healthy, contains 3 g of fat").

Authorized nutrient content claims relating to the content of calories, fat, saturated fat, cholesterol, sodium, and sugars are listed in Subpart D of Part 101. Requirements for "lean," "good source," "high," "more," "high potency," "antioxidant," "modified," and "fibre" claims are described in 101.13. Only the terms laid down may be used in making the claim; however, reasonable and non-misleading variations of spelling are acceptable, e.g. "hi," and "lo."

For substances that do not have a DV established, such as trans fats, only a factual statement that does not characterize the amount may be made. For example, "0 g trans fat" is an acceptable claim, but "trans fat free" or "low in trans fats" are not.

21 CFR Part 101.54 lays down requirements for relative nutrient content claims. A relative claim is a statement to compare the level of a nutrient in a food with the level in an appropriate reference food.

Relative claims include "light/reduced/added/fortified/enriched/extra/plus claims," for which the reference food must be a similar food.

For foods bearing relative claims, the following information is required:

1. the reference food must be identified and the percentage (or fraction) of nutrient difference must be declared (e.g. 15% more vitamin D than [reference food])." Such information must be declared next to the most prominent claim; and
2. quantitative information, comparing the level of a nutrient in the product per labeled serving with that in the reference food, next to most prominent claim or on the information panel.

For "More," "Fortified," "Plus," "Extra," "Enriched," and "Added" claims, there must be at least 10% more of the RDI (for vitamins or minerals) per Reference Amount Customarily Consumed (RACC) than the reference food and, where the nutrient has been added, fortification must be in accordance with FDA policy.

The RACCs listed under 21 CFR 101.12 for determining serving size for nutrition labelling are also used for eligibility for nutrition claims.

The regulations also define implied nutrient content claims which include the term "healthy."

To be labelled as "healthy":

- A food must meet the definition of "low" for fat and saturated fat
- Neither cholesterol nor sodium may be present at a level exceeding specified levels
- Contain at least ten percent of the RDI or DRV per RACC of vitamins A and C, calcium, iron, or protein
- Food must comply with definitions and declaration requirements for any specific nutrient content claims.

Information under "Nutrient Profiles" should also be considered.

5.6 Voluntary labelling statements: health claims

A "health claim" by definition has two essential components:

- a substance (whether a food, food component, or dietary ingredient) and
- a disease or health-related condition.

If a statement lacks either of these components then it does not qualify as a health claim.

There are three ways by which FDA determines which health claims may be used on a label or in labelling for a food or dietary supplement.

NLEA mandates FDA to issue regulations authorizing health claims for foods and dietary supplements after review of the scientific evidence submitted in health claim petitions. These claims are referred to as "Significant Scientific Agreement" (SSA) claims and are listed in 21 CFR 101 Subpart E. Examples of such claims include those linking fruits, vegetables, and grain products that contain fibre, particularly soluble fibre, and risk of coronary heart disease, and folate and neural tube defects.

The 1997 Food and Drug Administration Modernization Act (FDAMA) provides for health claims based on an authoritative statement of a scientific body of the US government or the National Academy of Sciences; such claims may be used after submission of a health claim notification to FDA for example, "Diets low in saturated fat, trans fat, and cholesterol may reduce the risk of heart disease." The conditions for this claim are that foods must be low in saturated fat and low in cholesterol, contain 0.5 g or less trans fat per reference amount, and less than 6.5 g total fat per reference amount (RACC).

The 2003 FDA *Consumer Health Information for Better Nutrition Initiative* provided for qualified health claims where the quality and strength of the scientific evidence falls below that required for FDA to issue an authorizing regulation. These "Qualified" health claims must be qualified to assure accuracy and non-misleading presentation to consumers. They must also undergo a review by FDA via a petition process. One example of this type claim is that for omega-3 fatty acids and coronary heart disease, which must be worded as follows:

"Supportive but not conclusive research shows that consumption of EPA and DHA omega-3 fatty acids may reduce the risk of coronary heart disease. One serving of [Name of the food] provides [] gram of EPA and DHA omega-3 fatty acids (see nutrition information for total fat, saturated fat, and cholesterol content)."

In addition, there are structure/function claims, which describe the role of a nutrient or dietary ingredient intended to affect normal structure or function in humans, for example "calcium builds strong bones." In addition, they may characterize the means by which a nutrient or dietary ingredient acts to maintain such structure or function, for example, "fibre maintains bowel regularity," or "antioxidants maintain cell integrity," or they may describe general well-being from consumption of a nutrient or dietary ingredient. Conventional food manufacturers are not required to notify FDA about these claims before placing on the market, and disclaimers are not required (as is the case for supplements). However, the manufacturer must provide scientific justification for use of any statements if questioned by the FDA.

There are nutrient profiles of a sort that apply to the making of health and nutrition claims laid down in the regulations. In order to prevent health claims from being used on foods of minimal nutritional value, FDA has adopted a policy known as the "jelly bean rule." Because jelly beans are normally low in saturated fat, and cholesterol, they might otherwise qualify to carry a health claim such as sodium and hypertension or saturated fat and coronary heart disease, but due to their low nutritive value this is inappropriate. Thus, 21 CFR Part 101.14 states that a food must provide at least ten percent of the DV of vitamin A, vitamin C, calcium, iron, fibre, or protein per RACC, without addition of any nutrients, in order to qualify for a health claim.

Also, *disqualifying nutrient levels* are levels of total fat, saturated fat, cholesterol, or sodium in a food above which the food will be disqualified from making a health claim. These levels are:

- **13.0** g of fat,
- **4.0** g of saturated fat,
- 60 mg of cholesterol,
- or 480 mg of sodium

per RACC, per label serving size, and only for foods with RACCs of 30 g or less, or two tablespoons or less, per 50 g.

5.7 Voluntary labelling statements: organic controls, GM labelling and claims

Since consumers are concerned about what a food contains, but even more about how it is manufactured, there is a growing body of regulation covering claims regarding methods of manufacture, such as for organically produced food products and those containing GM ingredients.

5.7.1 *Organic controls*

Title 7 of the Code of Federal Regulations (7 CFR) Part 205 details the final rule on the National Organic Program (NOP). Although the NOP is administered by USDA, its provisions are relevant to all types of food. This regulation lays down provisions on the labelling of organic products. Products may be labelled "100% Organic," "Organic," or "made with organic ingredients":

- "100% Organic"
- Foods labelled in this way must contain only organically produced ingredients.
- "Organic"
  Products labelled "Organic" must consist of at least 95% organically produced ingredients. Any remaining product ingredients must consist of non-agricultural substances approved on the National List (of ingredients permitted in organic products), or non-organically produced agricultural products that are not commercially available in organic form.
- "Made with organic ingredients"
  These are processed products that contain at least 70% organic ingredients. They may list up to three of the organic ingredients or food groups on the PDP.

Products containing less than 70% organic ingredients and products that are not processed by a certified organic handling operation may only disclose organic content in the list of ingredients inconspicuously, and may not display the US Organic Seal.

There is no threshold for GM-derived ingredients, unlike in the European Union, and use of bioengineering, or products thereof, is strictly prohibited.

The USA and EU signed an equivalency agreement on 15 February 2012, which became effective from June 2012 (USDA, 2012). Provided the terms are met, organic products certified to either USDA or EU organic standards may be sold and labelled as organic in both countries. Certification by accredited agents or bodies under either regime is recognized. Certain labelling statements, such as "made with organic X," as above, must be modified for the EU market.

5.7.2 GM labelling and claims

The FDA "Statement of Policy: Foods derived from New Plant Varieties" in 1992 did not establish special labelling requirements for bioengineered foods as a class; however, labelling requirements that apply to food in general also apply to foods produced using biotechnology.

These include Section 403(i) of the FD&C Act, which requires each food to bear a common or usual name, and Section 201(n), which requires the label of the food to reveal all material facts about the food.

If a bioengineered food significantly differs from its traditional counterpart, such that the common or usual name no longer adequately describes the new food, the name must be changed to describe the difference.

If an issue exists for the food or a constituent of the food regarding how the food is used or consequences of use, a statement must be made on the label to describe the issue.

If a bioengineered food has a significantly different nutritional property, its label must reflect the difference.

If a new food includes an allergen that consumers would not expect to be present based on the name of the food, the presence of the food must be declared on the label.

In 2001, FDA issued draft guidance on labelling that indicates whether foods have or have not been developed using bioengineering. This guidance explains how the provisions of the Federal Food, Drug & Cosmetic Act regarding labelling are applied to products of biotechnology by FDA, and gives examples of some statements that might be used in labelling, with a discussion accompanying each example, intended to provide guidance as to which statements can be made without being misleading.

This directly addressed use of the term "GM" and "GMO free" and concluded that consumers did not understand these acronyms (based upon consumer focus groups, at that time). Terms including the word "modified" are not technically accurate, unless they are in the context of bioengineering technology. In particular, FDA stated that the term "GMO free" may be misleading on most foods, because most foods do not contain organisms (seeds and foods like yogurt that contain microorganisms are exceptions). It would likely be misleading to suggest that a food that ordinarily would not contain entire "organisms" is "organism free."

There is also potential for the term "free" in a claim for absence of bioengineering to be inaccurate, because of the potential for adventitious presence of bioengineered material. FDA stated that the accuracy of the term "free" can only be ensured when there is a definition or threshold above which the term could not be used. FDA suggested that the term "free" either not be used in bioengineering label statements, or that it be used in a context that makes clear that a zero level of bioengineered material is not implied. Instead it suggested using statements where appropriate such as the following examples:

> "We do not use ingredients that were produced using biotechnology;"
> "This oil is made from soybeans that were not genetically engineered;" or
> "Our tomato growers do not plant seeds developed using biotechnology" (FDA, 2001).

Labelling must also not imply that a bioengineered ingredient is inferior or less safe because it is not recognized as being materially different from a non-engineered ingredient.

FDA's guidance also gave examples of statements that could be used if manufacturers wished to make statements about bioengineered ingredients used in their products.

In more recent years, consumer awareness and concern over GM foods has grown in the USA. Notably, a GM labelling bill, Proposition 37, was defeated in a referendum vote in November 2012. Connecticut and Maine have passed GM labelling laws, but both will apply only if neighboring states pass similar initiatives. Other states, including Washington and Vermont, have GMO bills under development at the time of writing. A Federal bill, the "Genetically Engineered Right-to-Know Act" was introduced in April 2013, but was defeated in the US Senate. FDA has pre-emption over state law, but states may adopt laws of their own that do not contain stricter requirements than those of Federal law. The impracticality of printing labels for sale in only certain states may mean that GM labelling, if required in even a few states, will become the norm nation-wide.

5.8 Voluntary labelling statements: gluten-free and "natural" claims

The gluten-free food market is going from strength to strength in the USA, driven not only by those with intolerance but by consumers with the general impression that avoidance of gluten is healthier. The same can be said of foods making "natural" claims. However, FDA takes a different approach to control these claims.

5.8.1 Gluten-free

On 5 August 2013, FDA published a Final Rule on gluten-free labelling (FDA, 2013). Gluten is defined as the proteins that occur naturally in wheat, rye, barley, and cross-breeds. The Final Rule requires that, in order to use the term "gluten-free," a food must meet all of the requirements of the definition, and contain less than 20 parts per million of gluten. A "gluten-free" food must not contain any of the following:

- an ingredient that is any type of wheat, rye, barley, or cross-breeds of these grains
- an ingredient derived from these grains and that has not been processed to remove gluten

- an ingredient derived from these grains that has been processed to remove gluten, if it results in the food containing 20 or more parts per million (ppm) gluten.

Foods that by their nature do not contain any gluten may also be labelled as "gluten-free," and "bottled spring water" is given as an example. Synonyms such as "free of gluten," "without gluten," and "no gluten" are considered equivalent, and if a similar phrase were used but the food did not comply, that food would be considered misbranded. Use of the claim on restaurant menus should comply with the same requirements.

Manufacturers have 1 year from the publication date to bring their labels into compliance. FDA also issued guidance on the new regulations; see Section 5.10.

5.8.2 "Natural" claims

The main written source of FDA's policy is in the preambles to the 1993 nutrient content claim regulations that implemented the NLEA Act and still represent current FDA policy on "natural." FDA's policy statement reads in part:

> *Because of resource limitations and other agency priorities, FDA is not undertaking rulemaking to establish a definition for 'natural' at this time. The agency will maintain its current policy (as discussed in the general principles proposal (56 FR 60421 at 60466)) not to restrict the use of the term 'natural' except for added color, synthetic substances, and flavors as provided in 101.22. Additionally, the agency will maintain its policy (Ref. 32) regarding the use of 'natural,' as meaning that nothing artificial or synthetic (including all color additives regardless of source) has been included in, or has been added to, a food that would not normally be expected to be in the food. Further, at this time the agency will continue to distinguish between natural and artificial flavors as outlined in 101.22.*
>
> *(FDA, 1993).*

The USDA's FSIS also has a definition for "natural," found in its Food Standards and Labeling Policy Handbook. This contains the same wording as FDA's definition, but also:

> *(2) the product and its ingredients are not more than minimally processed.*
>
> *Minimal processing may include: (a) those traditional processes used to make food edible or to preserve it or to make it safe for human consumption, e.g., smoking, roasting, freezing, drying, and fermenting, or (b) those physical processes which do not fundamentally alter the raw product and/or which only separate a whole, intact food into component parts, e.g., grinding meat, separating eggs into albumen and yolk, and pressing fruits to produce juices.Relatively severe processes, e.g., solvent extraction, acid hydrolysis, and chemical bleaching would clearly be considered more than minimal processing.*
>
> *(USDA, 2005).*

The FSIS policy also provides that all products claiming "natural" should be accompanied by a brief statement giving context, i.e., that the product is a natural food because it contains no artificial ingredients and is only minimally processed. This statement must appear either directly below or beside the claim, or be signposted to a location on the PDP.

FDA considers "natural" on a case-by-case basis and its definition is not process-based. It also does not address genetic engineering of ingredients. This has led to inconsistencies in labelling practice, and in 2011 a class action lawsuit was filed in California against Frito-Lay for labelling their Sun Chips and Tostitos products as "All Natural," although they were made with GM vegetable oils and maize (Julie Gengo v. Frito-Lay North America Inc.). A second lawsuit was filed in New York (Shake *et al.* v. Frito-Lay North America Inc.). The basis of these and similar suits is that consumers pay a premium for products promoted as "natural" and cannot be expected to understand products of genetic engineering as being "natural." Many of these lawsuits end in settlements by the defendants and an agreement to change their labelling. For example, Pepsico agreed to pay $9 million in a class action suit over "All Natural" labelling of its "Naked Juice" brand, over use of synthetic vitamins, synthetic fibres and claiming the use of "Non-GMO" ingredients when soya used in some products was alleged to be GM (Pappas v. Naked Juice Co). Pepsico also agreed to remove this labelling pending a decision from FDA.

FDA has been petitioned several times to adopt regulations to further define "natural" but has so far denied these requests. In a response letter to one such petition by The Sugar Association that requested a policy-based definition, FDA also noted, "A regulation defining the term 'natural' for food labeling is unnecessary because FDA's current policy is clear, well established, and reasonable. FDA has full authority to enforce its policy without issuing a regulation. FDA has a duty under section 403(a)(1) of the Federal Food, Drug, and Cosmetic Act to take action against any false or misleading food labeling" (FDA, 2006).

The issue has come to a head in another class action lawsuit, against Gruma Corporation, over representation of their Mission Tortilla Chips as "all natural" when they contain GM corn and corn derivatives. The judge in this case ordered a stay on further action in the case for 6 months to allow FDA to determine whether products containing GM-derived ingredients may be labelled as "natural" or "all natural." This order confirms the judge's view that this matter is in Federal, and not state, jurisdiction, but FDA is not obliged to issue any decision (Cox v. Gruma Corporation). At the time of writing, the outcome of this case, and the adoption of a formal definition, remain to be seen.

5.9 Future trends

As noted in Chapter 2, "The Food Labelling Modernization Act of 2013" was proposed, to update the food labelling provisions of the FD&C Act, many of which have been unchanged since 1938 (Congressional Fact Sheet, 19 September 2013).

Regarding labelling addressed in Chapter 5, the bill directs the Secretary of Health and Human Services (HHS) to:

- establish a single, standard front-of-pack labeling system in a timely manner for all food products required to bear nutrition labeling,

- update the definition of the term "healthy" according to the "Dietary Guidelines for Americans,"
- establish a standard definition for the term "natural,"
- require products claiming "made with whole grain" to conspicuously declare the percentage of whole grain,
- revise the Nutritional Facts panel to include the percent DVs for calories and sugar, as well as the amount of sugar that is not naturally occurring, ensure that "any product containing an amount of food reasonably consumed on a single occasion to state on the label that a single package contains one serving size."

At the time of writing, the bill was referred to the House Committee on Energy and Commerce and an identical bill, introduced in the Senate in November 2013, was submitted to the Committee on Health, Education, Labor, and Pensions. The likelihood of its eventual adoption in its current form is uncertain, but this bill could herald a new era in informing US consumers about their food choices.

In the meantime, however, FDA issued two proposed rules in March 2014: *Food Labeling: Revision of the Nutrition and Supplement Facts Label*, which includes provisions based on new nutrition science and design changes, and *Serving Sizes of Foods that can Reasonably be Consumed at One-Eating Occasion; Dual-Column Labeling; Updating, Modifying, and Establishing Certain Reference Amounts Customarily Consumed; Serving Size for Breath Mints; and Technical Amendments*, which addresses revised serving size requirements, criteria for labelling based on package size, and other issues.

Proposed changes based on new nutrition science include:

- Requiring the declaration of "Added Sugars" on the label,
- Removal of the requirement for declaring "Calories from fat,"
- Revision of nutrients of public health significance which must be declared on the label from vitamins A,C, D, and iron to calcium, vitamin D, potassium, and iron,
- Revision of Daily Values for certain nutrients, and change of units for certain nutrients and requirement to declare the amounts of vitamins and minerals and not only "% DV."

FDA is also proposing to amend the RACCs, as they, and therefore the nutrition values per serving size declared on the label, do not always reflect the current diet and eating habits of the US consumer. Twenty-five new RACCs would also be added. Rules for the calculation of serving size by package would also be revised, to more realistically reflect what is considered a "single serving."

There are also several changes to the current Nutrition Facts label proposed, to reflect aspects of the label deemed important in addressing current public health problems, such as obesity. The major alterations include:

- Highlighting of the caloric content of foods by increasing the type size and bolding the number of calories and servings per container.
- Moving % DV to the left of the label and declaring the actual amount, in addition to % DV, for all vitamins and minerals,
- Changing "Amount Per Serving" to "Amount per ___," with the blank filled in by the serving size in common household measurements.

- Replacing "Total Carbohydrate" with "Total Carbs" and indenting "Added Sugars" directly beneath the listing for "Sugars."
- Removing the existing footnote that describes the DVs for 2000 and 2500 calories to provide more space to better explain the percent dietary value.

The Comment Period for these Proposed Rules was open until 1 August, 2014. Once published as a Final Rule, FDA intends to allow 2 years from the effective date for compliance. Full details of the Proposed Rules can be found on FDA's website under the links provided under "Sources."

5.10 Sources of further information and advice

The following are useful resources for US food information recommended by the author:

Chapter IV of the Federal FD&C Act, which covers food, can be accessed online at: http://www.fda.gov/RegulatoryInformation/Legislation/FederalFoodDrugandCosmeticActFDCAct/FDCActChapterIVFood/default.htm

FDA has a wealth of documents and guidance available on their "Labeling and Nutrition" page, including Nutrition Labeling, Front-of-Pack Labeling, Allergens and Claims: http://www.fda.gov/Food/IngredientsPackagingLabeling/LabelingNutrition/default.htm, and the Labeling & Nutrition Guidance Documents & Regulatory Information: Guidance for Industry page contains links to further, detailed guidance: http://www.fda.gov/Food/GuidanceRegulation/GuidanceDocumentsRegulatoryInformation/LabelingNutrition/default.htm

FDA's A Food Labeling Guide, Revised January 2013 is found on the above page but deserves particular mention. This guide, presented in a question-and-answer format by subject area is a more user-friendly summarization of the major labelling requirements. www.fda.gov/FoodLabelingGuide

FDA's Compliance Policy Guides contain guidance on food labelling under Section 562: http://www.fda.gov/ICECI/ComplianceManuals/CompliancePolicyGuidanceManual/ucm119201.htm#SubChapter562

FDA's page on Topic-Specific Labeling Information contains details of areas which could not be fully explored in this Chapter: http://www.fda.gov/Food/GuidanceRegulation/GuidanceDocumentsRegulatoryInformation/LabelingNutrition/ucm2006864.htm

FDA, Questions and Answers: Gluten-Free Food Labeling Final Rule: http://www.fda.gov/Food/GuidanceRegulation/GuidanceDocumentsRegulatoryInformation/Allergens/ucm362880.htm

FDA's Warning Letters can be a useful source of guidance and interpretation: http://www.fda.gov/iceci/enforcementactions/WarningLetters/default.htm

FDA's Information Pages regarding the two Proposed Rules to amend Nutrition Facts Labeling has a wealth of guidance, links to the publications, and examples of the modified Nutrition Facts panel: http://www.fda.gov/Food/GuidanceRegulation/GuidanceDocumentsRegulatoryInformation/LabelingNutrition/ucm385663.htm

NIST Uniform Open Dating Regulation: National Institute of Standards and Technology (NIST Handbook No. 130, 2014 Ed.): http://www.nist.gov/pml/wmd/pubs/upload/17-section-IVe-14-h130-final.pdf

NOP: International Trade Policies: European Union. This page details the equivalency agreement between the USA and the EU: http://www.ams.usda.gov/NOPTradeEuropeanUnion

The Electronic Code of Federal Regulations, although not an official source, is an updated, online version of the Code of Federal Regulations and can be browsed by Title: http://www.ecfr.gov/

The Bill, Food Labeling Modernization Act of 2013 (H.R. 3147) can be accessed on the Library of Congress's Congress.gov website, at: http://beta.congress.gov/bill/113th-congress/house-bill/3147/text

TTB Ruling No. 2013–2, Voluntary Nutrient Content Statements in the Labeling and Advertising of Wines, Distilled Spirits, and Malt Beverages: http://www.ttb.gov/rulings/2013-2.pdf

USDA's FSIS page on Labeling and Label Approval give an extensive overview on policies, procedures, and guidance relevant to USDA-regulated products. The "Food Standards and Labeling Policy Book," which contains USDA's "Natural" policy definition is available under "Labeling Policies": http://www.fsis.usda.gov/wps/portal/fsis/topics/regulatory-compliance/labeling

References

Congressional Fact Sheet, The Food Labeling Modernization Act of 2013, Congress of the United States, 19.9.2013.

Cox v. Gruma Corporation, No. 12–6502, filed 26.12.2012, U.S. District Court for Northern District of California.

FDA (1993), 'Food Labeling Regulations Implementing the Nutrition Labeling and Education Act of 1990', *Federal Register* **58**(3), 2066–2941.

FDA (2013), Investigations Operations Manual, Exhibit 3-1, p. 99.

FDA, Guidance for Industry: Voluntary Labeling Indicating Whether Foods Have or Have Not Been Developed Using Bioengineering; Draft Guidance, January 2001.

Julie Gengo v. Frito Lay North America Inc, 2:11-cv-10322 in California Central District Court.

Shake *et al* v. Frito-Lay North America Inc and Pepsico Inc, in the U.S. District Court for the Eastern District of New York, No. 12–408. State of California, California menu labeling law (SB 1420).

The Code of Federal Regulations, Titles 21, 9, 27, 16, 19. US Government Printing Office, 2013. Available online at: http://www.ecfr.gov (Accessed 27 September 2013).

The Sugar Association, Citizen Petition re Definition of the Term 'Natural' for making claims on foods and beverages regulated by the FDA, 26.2.2006.

Health claims on food and beverage labels: comparing approaches in the EU and the USA

6

N. Baldwin
Intertek Scientific & Regulatory Consultancy, Farnborough, UK

6.1 Introduction

Within the food industry over the last couple of years, perhaps the most significant regulatory issue related to labelling has been that of health claims. More than ever before, with an ageing population and the pressures to achieve as children and adults, people want to get healthy and stay healthy. The ever-increasing demand for information, coupled with the obvious commercial opportunities for so-called "functional foods," has led to a plethora of products bearing health claims and, alas, many have not been based on sound science. Consumer and political concern has resulted in new stringent legislation in the European Union (EU), and most other jurisdictions are looking on and perhaps preparing for more of the same. Even the United States (USA), which has always been a more liberal market and by nature supportive of the "light-touch" regulatory approach, has its lines in the sand beyond which they regulate with a strong science-based submission process.

In this chapter, we will consider firstly the EU and the much slated and feted Nutrition and Health Claims Regulation, and how it interacts with the other relevant labelling legislation to provide the complete regulatory picture for the labelling of claims. This includes nutrition labelling and food information to consumers as well as foods for particular uses (PARNUTS), such as infant formula and the addition of vitamins, minerals and certain other substances to foods. We will then discuss how this compares with the more liberal market of the USA, where "dietary supplements" are regulated separately, and how enforcement discretion is of paramount importance. Finally, we will look at what the future holds for both jurisdictions, and whether they are moving closer together or further apart.

Advances in Food and Beverage Labelling. http://dx.doi.org/10.1533/9781782420934.2.87

6.2 Regulation of health and health-related claims in the EU

In the EU, the wording of nutrition and health claims for most foods have been regulated under on piece of legislation since 2006 (although foods for specific groups are considered on a case by case basis). The quantitative "triggers" for most of the claims involving vitamins and minerals and the main macronutrients come from reference values laid down under nutritional labelling legislation.

6.2.1 The Nutrition and Health Claims Regulation (NHCR) explained

Regulation (EC) No 1924/2006 on nutrition and health claims on foods is a landmark piece of EU legislation. Being a regulation, it applies directly in all Member States rather than being enacted into their legislation, and this is the case for most new EU legislation in order to avoid delays in implementation and indeed "selective translation," as may have been the case in the past. Under this regulation, nutrition and health claims are defined respectively in Article 2 as follows:

> *"nutrition claim" means any claim which states, suggests or implies that a food has particular beneficial nutritional properties due to:*

(a) the energy (calorific value) it
- **(i)** provides;
- **(ii)** provides at a reduced or increased rate; or
- **(iii)** does not provide; and/or

(b) the nutrients or other substances it
- **(i)** contains;
- **(ii)** contains in reduced or increased proportions; or
- **(iii)** does not contain;

> *"health claim" means any claim that states, suggests or implies that a relationship exists between a food category, a food or one of its constituents and health.*

The permitted list of nutrition claims is laid down under the Annex of the Regulation and was agreed by Member States based on the scientific advice of the European Food Safety Authority (EFSA) via Commission Regulation (EU) No 116/2010 of 9 February 2010 and Commission Regulation (EU) No 1047/2012 of 8 November 2012 amending Regulation (EC) No 1924/2006 of the European Parliament and of the Council with regard to the list of nutrition claims.

EFSA's main role was to decide what a recommended or adequate daily intake is for any given nutrient, and typically 15% of that amount recommends "source" and 30% represents "good source." The Annex of nutrition claims is completely generic.

In the case of health claims, things are slightly more complex because they are divided into essentially three types of claims:

1. Article 13 covers health claims describing or referring to:
   - the role of a nutrient or other substance in growth, development and the functions of the body; or
   - psychological and behavioural functions; or
   - without prejudice to Directive 96/8/EC, slimming or weight control or a reduction in the sense of hunger or an increase in the sense of satiety or to the reduction of the available energy from the diet.
2. Article 14 covers the other two types of:
   - claims referring to the reduction of disease risk, and
   - claims referring to children's development and health.

All of these claims require a submission to a Member State followed by the evaluation of an extensive scientific dossier to very stringent EFSA guidelines. EFSA are not known for their leniency, and the evaluation of the data has been likened to that of a new medicine, with the exception that, unlike medicines, a risk–benefit analysis is not conducted because safety and efficacy are not conducted concurrently on foods. In most cases, it is a food constituent/ingredient that is assessed rather than the food itself – another difference from medicines, which are evaluated as finished products. The European Commission maintains a register of approved claims, which is the easiest way to do an up-to-date check of which are authorised or not authorised (http://ec.europa.eu/nuhclaims/). The register is fully searchable so that, for example, if you looked for all claims that have been submitted for "biotin" and "brain," you would see Tables 6.1 and 6.2, respectively, as below.

For all three types of claims, submissions based on proprietary data are permitted. If it is ruled that the proprietary data (usually this means data unpublished at the time of submission) are pivotal to the scientific validity of the claim in EFSA's eyes, then the claim will be passed as only permitted to be used by or with the permission of the applicant for a period of 5 years, thereby giving a unique labelling advantage. Examples of this type of claim include water-soluble tomato concentrate and slowly digestible starch, which are summarised in Table 6.3 below.

Aside from the specific claim wording, all products containing health claims must adhere to the general rules that are laid down under Articles 3 and 10 of Regulation (EC) 1924/2006, recently reinforced under 2013/63/EU: *Commission Implementing Decision of 24 January 2013 adopting guidelines for the implementation of specific conditions for health claims laid down in Article 10 of Regulation (EC) No 1924/2006 of the European Parliament and of the Council.*

Article 3 states that

> *...health claims shall not*
>
> *(a) be false, ambiguous or misleading;*
>
> *(b) give rise to doubt about the safety and/or the nutritional adequacy of other foods;*
>
> *(c) encourage or condone excess consumption of a food;*
>
> *(d) state, suggest or imply that a balanced and varied diet cannot provide appropriate quantities of nutrients in general. Derogations in the case of nutrients for which sufficient quantities cannot be provided by a balanced and varied diet, including the conditions for their application, may be adopted in accordance with the*

Table 6.1 **Example entry from European Commission register when searching for all authorised claims submitted for "biotin"**

Claim type	Nutrient, substance, food or food category	Claim	Conditions of use of the claim/restrictions of use/reasons for non-authorisation	Health relationship	EFSA opinion reference/Journal reference	Commission Regulation	Status	Entry ID
Art.13(1)	Biotin	Biotin contributes to normal energy-yielding metabolism	The claim may be used only for food which is at least a source of biotin as referred to in the claim SOURCE OF [NAME OF VITAMIN/S] AND/OR [NAME OF MINERAL/S] as listed in the Annex to Regulation (EC) No 1924/2006.	Energy-yielding metabolism	2009;7(9):1209	Commission Regulation (EU) 432/2012 of 16/05/2012	Authorised	114, 117
Art.13(1)	Biotin	Biotin contributes to normal functioning of the nervous system	The claim may be used only for food which is at least a source of biotin as referred to in the claim SOURCE OF [NAME OF VITAMIN/S] AND/OR [NAME OF MINERAL/S] as listed in the Annex to Regulation (EC) No 1924/2006.	Function of the nervous system	2009;7(9):1209	Commission Regulation (EU) 432/2012 of 16/05/2012	Authorised	116

Art.13(1)	Biotin	Biotin contributes to normal macronutrient metabolism	The claim may be used only for food which is at least a source of biotin as referred to in the claim SOURCE OF [NAME OF VITAMIN/S] AND/OR [NAME OF MINERAL/S] as listed in the Annex to Regulation (EC) No 1924/2006.	Contribution to normal macronutrient metabolism	2009;7(9):1209, 2010;8(10):1728	Commission Regulation (EU) 432/2012 of 16/05/2012	Authorised	113, 114, 117, 4661
Art.13(1)	Biotin	Biotin contributes to normal psychological function	The claim may be used only for food which is at least a source of biotin as referred to in the claim SOURCE OF [NAME OF VITAMIN/S] AND/OR [NAME OF MINERAL/S] as listed in the Annex to Regulation (EC) No 1924/2006.	Contribution to normal psychological functions	2010;8(10):1728	Commission Regulation (EU) 432/2012 of 16/05/2012	Authorised	120
Art.13(1)	Biotin	Biotin contributes to the maintenance of normal hair	The claim may be used only for food which is at least a source of biotin as referred to in the claim SOURCE OF [NAME OF VITAMIN/S] AND/OR [NAME OF MINERAL/S] as listed in the Annex to Regulation (EC) No 1924/2006.	Maintenance of normal hair	2009;7(9):1209, 2010;8(10):1728	Commission Regulation (EU) 432/2012 of 16/05/2012	Authorised	118, 121, 2876

Continued

Table 6.1 **Continued**

Claim type	Nutrient, substance, food or food category	Claim	Conditions of use of the claim/restrictions of use/reasons for non-authorisation	Health relationship	EFSA opinion reference/Journal reference	Commission Regulation	Status	Entry ID
Art.13(1)	Biotin	Biotin contributes to the maintenance of normal mucous membranes	The claim may be used only for food which is at least a source of biotin as referred to in the claim SOURCE OF [NAME OF VITAMIN/S] AND/OR [NAME OF MINERAL/S] as listed in the Annex to Regulation (EC) No 1924/2006.	Maintenance of normal skin and mucous membranes	2009;7(9):1209	Commission Regulation (EU) 432/2012 of 16/05/2012	Authorised	115
Art.13(1)	Biotin	Biotin contributes to the maintenance of normal skin	The claim may be used only for food which is at least a source of biotin as referred to in the claim SOURCE OF [NAME OF VITAMIN/S] AND/OR [NAME OF MINERAL/S] as listed in the Annex to Regulation (EC) No 1924/2006.	Maintenance of normal skin and mucous membranes	2009;7(9):1209, 2010;8(10):1728	Commission Regulation (EU) 432/2012 of 16/05/2012	Authorised	115, 121

Art.13(1)	Biotin	Ensures normal functioning of the body's organ tissues and systems, particularly in old age – Stimulates physical work capacities – Recommended in case of intense fatigue and during recovery	Non-compliance with the regulation because on the basis of the scientific evidence assessed, this claimed effect for this food has not been substantiated.	Reduction of tiredness and fatigue	2010;8(10):1728		Non-authorised	119

Table 6.2 **Example entry from European Commission register when searching for all authorised claims submitted for "brain"**

Claim type	Nutrient, substance, food or food category	Claim	Conditions of use of the claim/restrictions of use/reasons for non-authorisation	Health relationship	EFSA opinion reference/journal reference	Commission regulation	Status	Entry ID
Art.13(1)	Docosahexa-enoic acid (DHA)	DHA contributes to maintenance of normal brain function	The claim may be used only for food which contains at least 40 mg of DHA per 100 g and per 100 kcal. In order to bear the claim, information shall be given to the consumer that the beneficial effect is obtained with a daily intake of 250 mg of DHA.	Maintenance of normal brain function	2010;8(10):1734, 2011;9(4):2078	Commission Regulation (EU) 432/2012 of 16/05/2012	Authorised	497, 501, 510, 513, 519, 521, 534, 540, 565, 626, 631, 688, 689, 690, 704, 742, 1323, 1360, 3148, 3151, 4294
Art.13(1)	Carbohy-drates	Carbohydrates contribute to the maintenance of normal brain function	In order to bear the claim, information shall be given to the consumer that the beneficial effect is obtained with a daily intake of 130 g of carbohydrates from all sources. The claim may be used for food which contains at least 20 g carbohydrates	Maintenance of normal brain function	2011;9(6):2226	Commission Regulation (EU) 1018/2013 of 23/10/2013	Authorised	603, 653

			which are metabolised by humans, excluding polyols, per quantified portion and complies with the nutrition claim LOW SUGARS or WITH NO ADDED SUGARS as listed in the Annex to Regulation (EC) No 1924/2006. The claim shall not be used on food which is 100% sugars.					
Art.14(1) (b)	Docosahexa-enoic acid (DHA)	Docosahexaenoic acid (DHA) maternal intake contributes to the normal brain development of the foetus and breastfed infants.	Information shall be given to pregnant and lactating women that the beneficial effect is obtained with a daily intake of 200 mg of DHA in addition to the recommended daily intake (RDI) for omega-3 fatty acids for adults, i.e., 250 mg DHA and eicosapentaenoic acid. The claim can be used only for food which provides a daily intake of at least 200 mg DHA.		2009;1007	Commission Regulation (EU) No 440/2011 of 0		

Table 6.3 **Examples of Article 13.5 health claims based on proprietary data which were given 5 years exclusivity**

Applicant – address	Nutrient, substance, food or food category	Claim	Conditions of use of the claim	Reference
Provexis Natural Products Ltd, Thames Court, 1 Victoria Street, Windsor, Berkshire, SL4 1YB, United Kingdom	Water-soluble tomato concentrate (WSTC) I and II	WSTC I and II helps maintain normal platelet aggregation, which contributes to healthy blood flow	Information to the consumer that the beneficial effect is obtained with a daily consumption of 3g WSTC I or 150 mg WSTC II in up to 250 ml of either fruit juices, flavoured drinks or yogurt drinks (unless heavily pasteurised).	Commission Decision of 17 December 2009 authorising a health claim on the effect of WSTC on platelet aggregation and granting the protection of proprietary data under Regulation (EC) No 1924/2006 of the European Parliament and of the Council (2009/980/EU)
Mondelēz International group, Three Parkway North Deerfield, IL 60015, United States of America	Slowly digestible starch (SDS)	Consumption of products high in SDS raises blood glucose concentration less after a meal compared to products low in SDS	The claim may be used only on food where the digestible carbohydrates provide at least 60% of the total energy and where at least 55% of those carbohydrates is digestible starch, of which at least 40% is SDS.	Commission Regulation (EU) No 851/2013 of 3 September 2013 authorising certain health claims made on foods, other than those referring to the reduction of disease risk and to children's development and health and amending Regulation (EU) No 432/2012

procedure referred to in Article 25(2), taking into account the special conditions present in Member States;

(e) refer to changes in bodily functions which could give rise to or exploit fear in the consumer, either textually or through pictorial, graphic or symbolic representations.

Article 10 states that

...health claims shall only be permitted if the following information is included in the labelling, or if no such labelling exists, in the presentation and advertising:

(a) a statement indicating the importance of a varied and balanced diet and a healthy lifestyle;

(b) the quantity of the food and pattern of consumption required to obtain the claimed beneficial effect;

(c) where appropriate, a statement addressed to persons who should avoid using the food; and

(d) an appropriate warning for products that are likely to present a health risk if consumed to excess.

Reference to general, non-specific benefits of the nutrient or food for overall good health or health-related well-being may only be made if accompanied by a specific health claim included in the lists provided for in Article 13 or 14.

So in summary, the Regulation (EC) 1924/2006 defines quite strictly when a health claim can be made, what the wording should be, and how the product is placed in the context of the whole diet. However, the regulation does overlap and coordinate with other EU labelling legislation.

6.2.2 How Regulation (EC) No 1924/2006 interacts with other legislation

When considering health claims labelling, it is important also to consider firstly the scope of Regulation (EC) 1924/2006. Article 1, Point 5 states:

This Regulation shall apply without prejudice to the following Community provisions:

(a) Directive 89/398/EEC and Directives adopted relating to foodstuffs for particular nutritional uses;

(b) Council Directive 80/777/EEC of 15 July 1980 on the approximation of the laws of the Member States relating to the exploitation and marketing of natural mineral waters;

(c) Council Directive 98/83/EC of 3 November 1998 on the quality of water intended for human consumption (3);

(d) Directive 2002/46/EC of the European Parliament and of the Council of 10 June 2002 on the approximation of the laws of the Member States relating to food supplements.

The term "without prejudice" means that where there are specific requirements laid down in these regulations and directives, they take precedence over the requirements of Regulation (EC) 1924/2006. So, for example, Point (a) refers to the current legislation on foods for particular nutritional uses. Directive 89/388/EEC has since been replaced by Directive 2009/39/EC of the European Parliament and of the Council of 6 May 2009 on foodstuffs intended for particular nutritional uses (recast), and this Directive in turn has been replaced by Regulation (EU) No 609/2013 of the European Parliament and of the Council of 12 June 2013 on food intended for infants and young children, food for special medical purposes, and total diet replacement for weight control and repealing Council Directive 92/52/EEC, Commission Directives 96/8/EC, 1999/21/EC, 2006/125/EC and 2006/141/EC, Directive 2009/39/EC of the European Parliament and of the Council and Commission Regulations (EC) No 41/2009 and (EC) No 953/2009. You can read these directives in more detail yourselves; however, at present, the most important message is that currently only infant formula itself has its own health claims procedure and list of approved claims; therefore, Regulation (EC) 1924/2006 does not apply to health claims related to infant formula. However, Regulation (EC) 1924/2006 (and specifically Article 14 claims for children's development and health) does directly apply to follow-on formula. Foods for special medical purposes have specific labels mandating who/which group of patients the foods are intended for, and so this aspect is also out of scope.

Under Point (d), food supplements have additional mandatory statements about doses and a healthy diet, etc., but apart from those stipulations, Regulation (EC) 1924/2006 applies in full.

The other major overlap is for claims related to the major nutrients, where the conditions of use for most vitamin and mineral claims simply refer to, for example:

"The claim may be used only for food which is at least a source of biotin as referred to in the claim SOURCE OF [NAME OF VITAMIN/S] AND/OR [NAME OF MINERAL/S] as listed in the Annex to Regulation (EC) No 1924/2006."

The Annex to Regulation (EC) 1924/2006 states:

> *SOURCE OF [NAME OF VITAMIN/S] AND/OR [NAME OF MINERAL/S]: A claim that a food is a source of vitamins and/or minerals, and any claim likely to have the same meaning for the consumer, may only be made where the product contains at least a significant amount as defined in the Annex to Directive 90/496/EEC or an amount provided for by derogations granted according to Article 6 of Regulation (EC) No 1925/2006 of the European Parliament and of the Council of 20 December 2006 on the addition of vitamins and minerals and of certain other substances to foods.*

So, we then go to Commission Directive 2008/100/EC of 28 October 2008 amending Council Directive 90/496/EEC on nutrition labelling for foodstuffs as regards to recommended daily allowances, energy conversion factors and definitions, which itself will be replaced by Regulation (EU) No 1169/2011 of the European Parliament and of the Council of 25 October 2011 on the provision of food information to consumers, amending Regulations (EC) No 1924/2006 and (EC) No 1925/2006 of the European Parliament and of the Council, and repealing Commission Directive 87/250/EEC, Council Directive 90/496/EEC, Commission Directive 1999/10/EC, Directive 2000/13/EC of the European Parliament and of the Council, Commission Directives 2002/67/EC and 2008/5/EC and Commission Regulation (EC) No 608/2004. Annex XIII, Point 2 states:

Significant amount of vitamins and minerals:

As a rule, the following values should be taken into consideration in deciding what constitutes a significant amount:

15% of the nutrient reference values specified in Point 1 supplied by 100 g or 100 mL in the case of products other than beverages,
7.5% of the nutrient reference values specified in Point 1 supplied by 100 mL in the case of beverages, or,
15% of the nutrient reference values specified in Point 1 per portion if the package contains only a single portion.

In terms of labelling and health claims in the EU, the most significant new activity is likely to be in relation to "generic descriptors." Under Regulation (EC) 1924/2006, Article 1, Point 4 states:

> *For generic descriptors (denominations), which have traditionally been used to indicate a particularity of a class of foods or beverages which could imply an effect on human health, a derogation from Paragraph 3 may be adopted in accordance with the procedure referred to in Article 25(2), on application by the food business operators concerned. The application shall be sent to the national competent authority of a Member State which will forward it to the Commission without delay. The Commission shall adopt and make public the rules for food business operators according to which such applications shall be made, so as to ensure that the application is dealt with transparently and within a reasonable time.*

The EU has recently adopted Commission Regulation (EU) No 907/2013 of 20 September 2013 setting the rules for applications concerning the use of generic descriptors. What this means is that companies and trade associations are preparing notifications to single and multiple Member States for a variety of descriptors, including "tonic water" and "probiotic." It remains to be seen how these will be received, firstly by the Member State(s) and secondly throughout the wider EU. But it does offer some potential hope for those who feel that the interpretation of the term "health claim" and its full submission requirements are over-strict.

The other ongoing issue to watch for in the future is how the EU decides eventually to handle health claims for "botanical" ingredients. These claims have been "parked" for some time whilst the European Commission reflects and considers whether the traditional use basis for many of these claims, rather than pure scientific data, may be used (if at all).

6.3 Regulation of health and health-related claims in the US

In the USA, health claims are split into three categories:

1. Health Claims
2. Qualified Health Claims
3. Structure Function Claims

6.3.1 Health claims

The definition of a health claim in the USA is laid down under 21 CFR 101.14(a)(1) (U.S. FDA, 2013a) "any claim made on the label or in labelling of a food, including a dietary supplement, that expressly or by implication, including 'third party' references, written statements (e.g., a brand name including a term such as 'heart'), symbols (e.g., a heart symbol), or vignettes, characterises the relationship of any substance *to a disease or health-related condition*. Implied health claims include those statements, symbols, vignettes, or other forms of communication that suggest, within the context in which they are presented, that a relationship exists between the presence or level of a substance in the food and a disease or health-related condition." This definition is closer to that of an Article 14 health claim in the EU, in that there is a link to disease risk reduction (but not treatment, cure or mitigation). In the EU, a risk factor is required for the claim, whereas in the USA direct disease risk-reduction claims are permitted.

All health claims must be approved by the US Food and Drug Administration (FDA) based on a critical review of publicly available scientific data. FDA has guidance for this under 21 CFR 101.9(k)(1), 101.14(c)-(d) and 21 CFR 101.70 (U.S. FDA, 2013a). It is based on the principle of "significant scientific agreement," and the scientific basis is very similar and no less stringent than that of the EU. An up-to-date list of the currently permitted approved health claims can be found on the FDA website (U.S. FDA, 2013b), wherein the terms of use for each claim are strictly specified. For example, 21 CFR 101.72 for the health claim related to calcium, vitamin D and osteoporosis (U.S. FDA, 2013a) states:

> *...(c) Requirements. (1) All requirements set forth in §101.14 shall be met.*
>
> *(2) Specific requirements – (i) Nature of the claim. A health claim associating calcium or, when appropriate, calcium and vitamin D with a reduced risk of osteoporosis may be made on the label or labelling of a food described in paragraphs (c)(2)(ii) and (d)(1) of this section, provided that:*
>
> *(A) The claim makes clear the importance of adequate calcium intake, or when appropriate, adequate calcium and vitamin D intake, throughout life, in a healthful diet, are essential to reduce osteoporosis risk. The claim does not imply that adequate calcium intake, or when appropriate, adequate calcium and vitamin D intake, is the only recognised risk factor for the development of osteoporosis;*
>
> *(B) The claim does not attribute any degree of reduction in risk of osteoporosis to maintaining an adequate dietary calcium intake, or when appropriate, an adequate dietary calcium and vitamin D intake, throughout life.*
>
> *(ii) Nature of the food. (A) The food shall meet or exceed the requirements for a 'high' level of calcium as defined in §101.54(b);*
>
> *(B) The calcium content of the product shall be assimilable;*
>
> *(C) Dietary supplements shall meet the United States Pharmacopeia (USP) standards for disintegration and dissolution applicable to their component calcium salts, except that dietary supplements for which no USP standards exist shall exhibit appropriate assimilability under the conditions of use stated on the product label;*
>
> *(D) A food or total daily recommended supplement intake shall not contain more phosphorus than calcium on a weight per weight basis.*

> *(d) Optional information. (1) The claim may include the term 'vitamin D' if the food meets or exceeds the requirements for a 'high' level of vitamin D as defined in §101.54(b);*
>
> *(2) The claim may include information from paragraphs (a) and (b) of this section.*
>
> *(3) The claim may make reference to physical activity.*
>
> *(4) The claim may include information on the number of people in the United States, including the number of people in certain subpopulations in the United States, who have osteoporosis or low bone density. The sources of this information must be identified, and it must be current information from the National Center for Health Statistics, the National Institutes of Health, or the National Osteoporosis Foundation.*
>
> *(5) The claim may state that the role of adequate calcium intake, or when appropriate, the role of adequate calcium and vitamin D intake, throughout life is linked to reduced risk of osteoporosis through the mechanism of optimising peak bone mass during adolescence and early adulthood. The phrase 'build and maintain good bone health' may be used to convey the concept of optimising peak bone mass. The claim may also state that adequate intake of calcium, or when appropriate, adequate intake of calcium and vitamin D, is linked to reduced risk of osteoporosis through the mechanism of slowing the rate of bone loss for persons with a family history of the disease, post-menopausal women, and elderly men and women.*
>
> *(e) Model health claims. The following model health claims may be used in food labelling to describe the relationship between calcium and osteoporosis:*
>
> *Adequate calcium throughout life, as part of a well-balanced diet, may reduce the risk of osteoporosis.*
>
> *Adequate calcium as part of a healthful diet, along with physical activity, may reduce the risk of osteoporosis in later life.*
>
> *(f) Model additional health claims for calcium and vitamin D. The following model health claims may be used in food labelling to describe the relationship between calcium, vitamin D, and osteoporosis:*
>
> *Adequate calcium and vitamin D throughout life, as part of a well-balanced diet, may reduce the risk of osteoporosis.*

Adequate calcium and vitamin D as part of a healthful diet, along with physical activity, may reduce the risk of osteoporosis in later life.

All requirements set forth in §101.14 shall be met. This includes generic disqualifying nutrient profiles, which are defined as "the levels of total fat, saturated fat, cholesterol, or sodium in a food above which the food will be disqualified from making a health claim." These levels are 13.0 g of fat, 4.0 g of saturated fat, 60 mg of cholesterol, or 480 mg of sodium, per reference amount customarily consumed, per label serving size, and, only for foods with reference amounts customarily consumed of 30 g or less or 2 tablespoons or less, per 50 g. For dehydrated foods that must have water added to them prior to typical consumption, the per 50 g criterion refers to the as-prepared form. Any one of the levels, on a per reference amount customarily consumed, a per label serving size or, when applicable, a per 50 g basis, will disqualify a food from making a health claim unless an exception is provided in subpart E of this part, except that:

1. *The levels for a meal product as defined in 101.13(l) are 26.0 g of fat, 8.0 g of saturated fat, 120 mg of cholesterol, or 960 mg of sodium per label serving size, and*

2. *The levels for a main dish product as defined in 101.13(m) are 19.5 g of fat, 6.0 g of saturated fat, 90 mg of cholesterol, or 720 mg of sodium per label serving size.*

So, each claim is evaluated using a case-by-case approach. In addition to the claim above, the following other "full" health claims that meet significant scientific agreement and that have been approved to date are related to:

- Dietary lipids (fat) and cancer
- Dietary saturated fat and cholesterol, and risk of coronary heart disease
- Dietary non-cariogenic carbohydrate sweeteners and dental caries
- Fibre-containing grain products, fruits and vegetables, and cancer
- Folic acid and neural tube defects
- Fruits and vegetables, and cancer
- Fruits, vegetables and grain products that contain fibre, particularly soluble fibre, and risk of coronary heart disease
- Sodium and hypertension
- Soluble fibre from certain foods and risk of coronary heart disease
- Soy protein and risk of coronary heart disease
- Stanols/sterols and risk of coronary heart disease.

Health claims are not permitted on infant formula or on medical foods (which are regulated as orphan drugs as well as foods). They are permitted in dietary supplements on a case-by-case basis. All health claims are generic; there is no procedure in the USA for exclusivity of health claims based on pivotal proprietary data as there are in the EU.

6.3.2 Qualified health claims

Qualified health claims are regulated in a similar way to full health claims that meet significant scientific agreement under "interim procedures" and "enforcement discretion" of the FDA. They are basically petitions made to the FDA based on the same criteria but requesting less definite wording that reflects the balance of the publicly available data. The current list of these claims is maintained on the FDA website (U.S. FDA, 2013c) and includes a variety of claims from:

> *Supportive but not conclusive research shows that consumption of eicosapentaenoic acid (EPA) and docosahexaenoic acid (DHA) omega-3 fatty acids may reduce the risk of coronary heart disease. One serving of [Name of the food] provides [] gram of EPA and DHA omega-3 fatty acids (U.S. FDA, 2013d),"* to:
>
> *One small study suggests that chromium picolinate may reduce the risk of insulin resistance, and therefore possibly may reduce the risk of type 2 diabetes. FDA concludes, however, that the existence of such a relationship between chromium picolinate and either insulin resistance or type 2 diabetes is highly uncertain (U.S. FDA, 2013e).*

Interestingly under this "interim procedure," the FDA has also published a letter of enforcement discretion for a claim related to 100% Whey-Protein Partially Hydrolyzed Infant Formula and Reduced Risk of Atopic Dermatitis (U.S. FDA, 2011).

6.3.3 Structure function claims

Structure function claims are similar in nature to the health claims defined under Article 13 in Regulation (EC) No 1924/2006. They are defined and permitted under the Dietary Supplement Health and Education Act of 1994 (DSHEA), Section 403(r)(6) and Section 101.93(f) to the Food Drug and Cosmetics (FD&C) Act:

> *Permitted structure/function statements. Dietary supplement labels or labelling may, subject to the requirements in paragraphs (a) through (e) of this section, bear statements that describe the role of a nutrient or dietary ingredient intended to affect the structure or function in humans or that characterise the documented mechanism by which a nutrient or dietary ingredient acts to maintain such structure or function, provided that such statements are not disease claims under paragraph (g) (21 CFR 101.93) (U.S. FDA, 2013a). ...Section 403(r)(6) of the FD&C Act does not apply to conventional foods, however structure/function claims may be made on a conventional food provided the effects are derived from the nutritive value of the food.*

Put simply, structure function claims relate to the maintenance and support of health. For example, "calcium helps to maintain healthy teeth and bones." As stated above, they are largely used for dietary supplement products, and whilst no formal submission procedure is involved, there is the automatic requirement of due diligence in that the totality of the scientific evidence is supportive and that the claim is truthful and not misleading. All dietary supplements producers, in addition, must notify the FDA that they are using the claim within 30 days of first marketing of a new product. There is no such requirement for food producers. FDA will confirm only that the product does not cross over into the definition of a drug claim. They do not routinely check in detail the specific science related to the claim itself. For dietary supplements, the claim must include a mandatory disclaimer statement that is provided for in the law. Section 403(r)(6) of the FD&C Act states (U.S. FDA, 2013a):

"This statement has not been evaluated by the FDA. This product is not intended to diagnose, treat, cure, or prevent any disease."

Structure function claims are by far the most widely used in the USA, probably because there is no formal submission procedure and regulation is largely based on enforcement discretion.

6.3.4 How the health claims regulations interact with other legislation

As for the EU, there is considerable overlap with other legislation for health-related claims. Mainly, the overlap involves "nutrient content claims." Nutrient content claims are similar to nutrition claims in the EU, in that they refer to claims on food products "that directly or by implication characterises the level of nutrient in a food." The criteria for such claims are laid down under the Code of Federal Regulations (CFR): 21 CFR 101.13(j)(2)(i) and 21 CFR 101.13(j)(2)(ii) (U.S. FDA, 2013a). This regulation provides very detailed instructions for their use and implementation. The best place to get a working knowledge of claims is the FDA guidance documents (U.S. FDA, 2013f). The principles are very similar to those of the EU; however, they are far more extensive, indeed extending even to the term "healthy" (which might be considered as a "generic descriptor" or even as an established trademark or "fancy name" under Regulation (EC) No 1924/2006), as can be seen in Tables 6.4 and 6.5, which are reproduced from the FDA website for convenience below.

Table 6.4 Permitted nutrient content claims in the United States

Claim	Requirements
Relative claims	
"Light"	(1) A food representative of the type of food bearing the claim (e.g., average value of top three brands or representative value from valid data base), and (2) Similar food (e.g., potato chips for potato chips)
"Reduced" and "Added" (or "Extra," "Plus," "Fortified" and "Enriched")	(1) An established regular product or average representative product, and (2) Similar food.
"More" and "Less" (or "Fewer")	(1) An established regular product or average representative product, and (2) A dissimilar food in the same product category which may be generally substituted for the labeled food (e.g., potato chips for pretzels) or a similar food.
Other nutrient content claims	
"High," "Rich In," or "Excellent Source of"	Contains 20% or more of the daily values (DV) per reference amount customarily consumed (RACC). May be used on meals or main dishes to indicate that the product contains a food that meets the definition, but may not be used to describe the meal. *21 CFR 101.54(b)*
"Good Source," "Contains," or "Provides"	10%-19% of the DV per RACC. These terms may be used on meals or main dishes to indicate that the product contains a food that meets the definition but may not be used to describe the meal. *21 CFR 101.54(e)*
"More," "Fortified," "Enriched," "Added," "Extra," or "Plus"	10% or more of the DV per RACC than an appropriate reference food. May only be used for vitamins, minerals, protein, dietary fiber and potassium. *21 CFR 101.54(e)*
"Lean"	On seafood or game meat products that contain less than 10 g total fat, 4.5 g or less saturated fat, and less than 95 mg cholesterol per RACC and per 100 g (for meals and main dishes, meets criteria per 100 g and per labeled serving). On mixed dishes not measurable with a cup (as defined in *21 CFR 101.12(b)* in Table 2) that contain less than 8 g total fat, 3.5 g or less saturated fat and less than 80 mg cholesterol per RACC. *21 CFR 101.62(e)(1)-(3)*

Table 6.4 Continued

Claim	Requirements
"Extra Lean"	On seafood or game meat products that contains less than 5 g total fat, less than 2 g saturated fat and less than 95 mg cholesterol per RACC and per 100 g (for meals and main dishes, meets criteria per 100 g and per labeled serving). *21 CFR 101.62(e)(4)* and *(5)*
"High Potency"	May be used on foods to describe individual vitamins or minerals that are present at 100% or more of the RDI per RACC or on a multi-ingredient food product that contains 100% or more of the RDI for at least 2/3 of the vitamins and minerals with RDIs and that are present in the product at 2% or more of the RDI (e.g., "High potency multivitamin, multimineral dietary supplement tablets"). *21 CFR101.54(f)*
"Modified"	May be used in statement of identity of a food that bears a relative claim (e.g., "Modified fat cheesecake contains 35% less fat than our regular cheesecake.") *21 CFR 101.13(k)*
"Fiber" Claims	If a fiber claim is made and the food is not low in total fat, then the label must disclose the level of total fat per labeled serving. *21 CFR 101.54(d)(1)*
Claims using the term "antioxidant"	For claims characterising the level of antioxidant nutrients in a food: 1. an RDI must be established for each of the nutrients that are the subject of the claim; 2. each nutrient must have existing scientific evidence of antioxidant activity; 3. the level of each nutrient must be sufficient to meet the definition for "high," "good source," or "more"; Beta-carotene may be the subject of an antioxidant claim when the level of vitamin A present as beta-carotene in the food is sufficient to qualify for the claim. 4. Name(s) of nutrient(s) that is (are) the subject of the claim is (are) included as part of the claim. (e.g., high in antioxident vitamins C and E) *21 CFR 101.54(g)*

Nutrient content claims can often be used in association with "authoritative body statements," such as the Institute of Medicine (IOM), to form complete structure function claims. For example:

"This product is a good source of calcium and vitamin D. Calcium and vitamin D are two essential nutrients long known for their role in bone health (IOM, 2010)."

Unlike the EU, the USA prohibits some nutrient content claims for children less than 2 years of age, specifically:

- Claims describing the percentage of vitamins and minerals in a food in relation to a daily value;
- Claims on infant formulas;
- The terms "Unsweetened" and "Unsalted" as taste claims;
- "Sugar Free" and "No Added Sugar" claims on dietary supplements only.

Table 6.5 Conditions for the use of "healthy"

	Individual food	Seafood/game meat	Meal/main dish
Total fat	low fat	<5 g fat /RACC and /100g	Low fat
Saturated fat	low sat fat	<2 g sat fat /RACC and/100 g	Low sat fat
Sodium	≤ 480 mg /RACC and /l.s.; or /50 g, if RACC is small	≤480 mg /RACC and /l.s.; or /50 g, if RACC is small	≤600 mg /l.s.
Cholesterol	≤ disclosure level	<95 mg /RACC and /100 g	≤90 mg /l.s.
Beneficial nutrients	Contains at least 10% of DV/ RACC for vitamins A, C, calcium, iron, protein, or fibre except: raw fruits and vegetables; or a single ingredient or mixture of frozen or canned single ingredient fruits and vegetables (may include ingredients whose addition does not change the nutrient profile of the fruit or vegetable); enriched cereal-grain products that conform to a standard of identity in *21 CFR 136, 137, or 139.*	Contains at least 10% of DV / RACC for vitamins A, C, calcium, iron, protein, or fibre	Contains at least 10% of the DV /l.s. of two nutrients (for a main dish product) or of three nutrients (for a meal product) of vitamin. A, vitamin. C, calcium, iron, protein, or fibre.
Fortification	Per *21 CFR 104.20*	Per *21 CFR 104.20*	Per *21 CFR 104.20*

NOTE: l.s. = labelled serving; RACC = reference amount customarily consumed per eating occasion; small RACC = 30 g or less, or two tablespoons or less.

As discussed above, health claims are not normally permitted for infant formula and special medical foods.

6.4 Summary, conclusions and future trends

The regulations of health claim labelling in the EU and the USA have, in both cases, their foundations in the principles of significant scientific agreement. Whilst in the EU, this consensus is defined by EFSA, which itself is made up of "staff" scientists and committees' consisting of independent academics, whilst in the USA, the review is done by the "staffers" of the FDA alone based on peer-reviewed publications.

Table 6.6 A comparison of Health Claim Regulations between EU and USA

Claim type/conditions	EU	US
New "maintenance"/"supports" type claims	Pre-submission required as a health claim.	No pre-submission required as a structure function claim. Notification is required for dietary supplements only; FDA simply to check the claim is not medicinal.
New disease risk-reduction claims	Pre-submission required as a health claim based on disease *risk factor* reduction.	Pre-submission required as a health claim. Direct disease risk-reduction claims are permitted.
New qualified health claims	Not excluded in principle under EU law, but in practice the EFSA is reluctant to deliver "qualified" opinions, such as "may" or "might."	Pre-submission required and subject to FDA "enforcement discretion."
New proprietary health claims	Pre-submission required. Exclusivity can be granted for 5 years based on unpublished pivotal scientific data.	Not permitted. All health claims must be based on published data.
Nutrient profiles	Not yet defined.	Defined at both a generic and a specific claim level.
New health claims made for infant formula and medical foods	Outside of scope.	Outside of scope of health claims, but can be considered under qualified health claims on a case-by-case basis.
Generic descriptors such as "healthy"	Regulated under new Member State notification procedures.	Some covered under nutrient content claims. The remainder under FDA enforcement discretion.

This means that, in fact, FDA arguably has the easier job to manage and obtain a consensus, as the difficulties in getting any group of academics to agree are widely recognised. Table 6.6 provides a summary of the major differences in regulatory definitions and the approval processes; it is clear that there are more similarities than differences.

However, in the USA, currently structure function claims are subject largely to self-regulation, which reflects the major difference in approach to regulation between the two jurisdictions. The threat and financial consequences of litigation in the USA are seen as far greater deterrents than in the EU. However, with the regulation of dietary supplements under DSHEA constantly being challenged and subject to review by consumer groups, the pharmaceuticals industry and other non-governmental organisations, may change in future. Whether or not the enforcement discretion and interim measures related to qualified health claims will become more permanent remains to be seen as, whilst many of the claims are actually too complicated for consumer understanding, they do represent the first steps towards recognition of disease risk relationships and encourage further research to "upgrade" them to full claims in future.

In the EU, whilst there is the regulatory expectation and the general will of the Member States to establish nutrient profiles, arguments between Member States have been so prolonged and difficult that even the European Commission has put this issue aside, in the same way as they have dealt with botanicals. We expect more movement on that in the coming months and years as we do over the new procedures for "generic descriptors."

Overall, from a practical perspective, harmonisation of approved claims would greatly benefit trade between the USA and EU, with the US supplements industry arguably having the greatest barriers to trade at present. What has been a relatively free marketing exchange from the US dietary supplement industry based on structure function claims to the EU now requires extensive re-labelling of most products, as enforcement is getting stronger in the EU now that the actual rules have been more clearly defined. This, of course, leads to the most important question, which is whether the real issue is not a health claims procedure but its enforcement. In the EU, the rules are now much clearer to support the enforcers, but the resources probably have not changed. So will clearer rules alone, with the same amount of enforcement, catch and deal with the core issue of misleading the consumer, especially with the internet such a powerful trans-border marketing platform? We will see as time progresses.

References

Commission Decision (2009/980/EU) of 17 December 2009 authorising a health claim on the effect of water-soluble tomato concentrate on platelet aggregation and granting the protection of proprietary data under Regulation (EC) No 1924/2006 of the European Parliament and of the Council (notified under document C(2009) 10113). *Off J Eur Union,* **52**(L336), 55–57. Available from: http://eur-lex.europa.eu/LexUriServ/LexUriServ.do?uri=OJ:L:2009:336:0055:0057:EN:PDF.

Commission Implementing Decision (2013/63/EU) of 24 January 2013 adopting guidelines for the implementation of specific conditions for health claims laid down in Article 10 of Regulation (EC) No 1924/2006 of the European Parliament and of the Council. *Off J Eur Union,* **56**(L22), 25–28. Available from: http://eur-lex.europa.eu/LexUriServ/LexUriServ.do?uri=OJ:L:2013:022:0025:0028:EN:PDF.

Commission Directive 2008/100/EC of 28 October 2008 amending Council Directive 90/496/EEC on nutrition labelling for foodstuffs as regards recommended daily allowances, energy conversion factors and definitions. *Off J Eur Union,* **51**(L285), 9–12. Available from: http://eur-lex.europa.eu/LexUriServ/LexUriServ.do?uri=OJ:L:2008:285:0009:0012:EN:PDF.

Commission Directive 2009/39/EC of the European Parliament and of the Council of 6 May 2009 on foodstuffs intended for particular nutritional uses. *Off J Eur Union,* **52**(L124), 21–29. Available from: http://eur-lex.europa.eu/LexUriServ/LexUriServ.do?uri=OJ:L:2009:124:0021:0029:EN:PDF.

Commission Regulation (EC) No 1924/2006 of the European Parliament and of the Council of 20 December 2006 on nutrition and health claims made on foods. *Off J Eur Union,* **49**(L404), 9–25. Available from: http://eur-lex.europa.eu/LexUriServ/LexUriServ.do?uri=CONSLEG:2006R1924:20100302:EN:PDF [Consolidated Version: 2010-03-02].

Commission Regulation (EU) No 116/2010 of 9 February 2010 amending Regulation (EC) No 1924/2006 of the European Parliament and of the Council with regard to the list of nutrition claims. *Off J Eur Union,* **53**(L37), 16–18. Available from: http://eur-lex.europa.eu/LexUriServ/LexUriServ.do?uri=OJ:L:2010:037:0016:0018:EN:PDF.

Commission Regulation (EU) No 440/2011 of 6 May 2011 on the authorisation and refusal of authorisation of certain health claims made on foods and referring to children's development and health. *Off J Eur Union,* **54**(L119), 4–9. Available from: http://eur-lex.europa.eu/LexUriServ/LexUriServ.do?uri=OJ:L:2011:119:0004:0009:EN:PDF.

Commission Regulation (EU) No 1169/2011 of the European Parliament and of the Council of 25 October 2011 on the provision of food information to consumers, amending Regulations (EC) No 1924/2006 and (EC) No 1925/2006 of the European Parliament and of the Council, and repealing Commission Directive 87/250/EEC, Council Directive 90/496/EEC, Commission Directive 1999/10/EC, Directive 2000/13/EC of the European Parliament and of the Council, Commission Directives 2002/67/EC and 2008/5/EC and Commission Regulation (EC) No 608/2004. *Off J Eur Union,* **54**(L304), 18–63. Available from: http://eur-lex.europa.eu/LexUriServ/LexUriServ.do?uri=OJ:L:2011:304:0018:0063:EN:PDF.

Commission Regulation (EU) No 432/2012 of 16 May 2012 establishing a list of permitted health claims made on foods, other than those referring to the reduction of disease risk and to children's development and health. *Off J Eur Union,* **55**(L136), 1–40. Available from: http://eur-lex.europa.eu/LexUriServ/LexUriServ.do?uri=OJ:L:2012:136:0001:0040:EN:PDF.

Commission Regulation (EU) No 1047/2012 of 8 November 2012 amending Regulation (EC) No 1924/2006 with regard to the list of nutrition claims. *Off J Eur Union,* **55**(L310), 36–37. Available from: http://eur-lex.europa.eu/LexUriServ/LexUriServ.do?uri=OJ:L:2012:310:0036:0037:EN:PDF.

Commission Regulation (EU) No 609/2013 of 12 June 2013 on food intended for infants and young children, food for special medical purposes, and total diet replacement for weight control and repealing Council Directive 92/52/EEC, Commission Directives 96/8/EC, 1999/21/EC, 2006/125/EC and 2006/141/EC, Directive 2009/39/EC of the European Parliament and of the Council and Commission Regulations (EC) No 41/2009 and (EC) No 953/2009. *Off J Eur Union,* **56**(L181), 35–56. Available from: http://new.eur-lex.europa.eu/legalcontent/EN/TXT/PDF/?uri=CELEX:32013R0609&from=EN.

Commission Regulation (EU) No 851/2013 of 3 September 2013 authorising certain health claims made on foods, other than those referring to the reduction of disease risk and to children's development and health and amending Regulation (EU) No 432/2012. *Off J Eur Union,* **56**(L235), 3–7. Available from: http://eur-lex.europa.eu/LexUriServ/LexUriServ.do?uri=OJ:L:2013:235:0003:0007:EN:PDF.

Commission Regulation (EU) No 907/2013 of 20 September 2013 setting the rules for applications concerning the use of generic descriptors (denominations). *Off J Eur Union,* **56**(L251), 7–9. Available from: http://eur-lex.europa.eu/LexUriServ/LexUriServ.do?uri=OJ:L:2013:251:0007:0009:EN:PDF.

Commission Regulation (EU) No 1018/2013 of 23 October 2013 amending Regulation (EU) No 432/2012 establishing a list of permitted health claims made on foods other than those referring to the reduction of disease risk and to children's development and health. *Off J Eur Union,* **56**(L282), 43–45. Available from: http://eur-lex.europa.eu/LexUriServ/LexUriServ.do?uri=OJ:L:2013:282:0043:0045:EN:PDF.

EFSA (2009), Scientific Opinion of the Panel on Dietetic Products, Nutrition and Allergies on a request from Merck Selbstmedikation GmbH on DHA and support of the cognitive development of the unborn child and breastfed infant [tuna oil] (question no EFSA-Q-2008-773, adopted on 13 March 2009, European Food Safety Authority). *EFSA J,* **1007**, 1–14. Available from: http://www.efsa.europa.eu/en/efsajournal/pub/1007.htm.

EFSA (2009), EFSA Panel on Dietetic Products, Nutrition and Allergies (NDA); Scientific opinion on the substantiation of health claims related to biotin and energy-yielding metabolism (ID 114, 117), macronutrient metabolism (ID 113, 114, 117), maintenance of skin and mucous membranes (ID 115), maintenance of hair (ID 118, 2876) and function of the nervous system (ID 116) pursuant to Article 13(1) of Regulation (EC) No 1924/2006 on request from the European Commission (question no EFSA-Q-2008-900, EFSA-Q-2008-901, EFSA-Q-2008-902, EFSA-Q-2008-903, EFSA-Q-2008-904, EFSA-Q-2008-905, EFSA-Q-2008-3609, adopted on 02 July 2009 by European Food Safety Authority). *EFSA J,* **7**(9), 1209. [17 pp.]. doi:10.2903/j.efsa.2009.1209. Available from: http://www.efsa.europa.eu/en/scdocs/scdoc/1209.htm.

EFSA (2010), Panel on Dietetic Products, Nutrition and Allergies (NDA); Scientific Opinion on the substantiation of health claims related to biotin and maintenance of normal skin and mucous membranes (ID 121), maintenance of normal hair (ID 121), maintenance of normal bone (ID 121), maintenance of normal teeth (ID 121), maintenance of normal nails (ID 121, 2877), reduction of tiredness and fatigue (ID 119), contribution to normal psychological functions (ID 120) and contribution to normal macronutrient metabolism (ID 4661) pursuant to Article 13(1) of Regulation (EC) No 1924/2006 (question No EFSA-Q-2008-908, EFSA-Q-2008-3610, adopted on 11 February 2010 and question No EFSA-Q-2008-906, EFSA-Q-2008-907, EFSA-Q-2010-00614, adopted on 09 July 2010 by European Food Safety Authority). *EFSA J,* **8**(10), 1728. [19 pp.]. doi:10.2903/j.efsa.2010.1728. Available from: http://www.efsa.europa.eu/en/efsajournal/pub/1728.htm.

EFSA (2010), EFSA Panel on Dietetic Products, Nutrition and Allergies (NDA); Scientific Opinion the substantiation of a health claim related to docosahexaenoic acid (DHA) and maintenance of normal (fasting) blood concentrations of triglycerides (ID 533, 691, 3150), protection of blood lipids from oxidative damage (ID 630), contribution to the maintenance or achievement of a normal body weight (ID 629), brain, eye and nerve development (ID 627, 689, 704, 742, 3148, 3151), maintenance of normal brain function (ID 565, 626, 631, 689, 690, 704, 742, 3148, 3151), maintenance of normal vision (ID 627, 632, 743, 3149) and maintenance of normal spermatozoa motility (ID 628) pursuant to Article 13(3) of Regulation (EC) no 1924/2006 (question no EFSA-Q-2008-1320, EFSA-Q-2008-1416, EFSA-Q-2008-1417, EFSA-Q-2008-1478, … *[etc.]*, adopted on 09 July 2010 by European Food Safety Authority). *EFSA J,* **8**(10), 1734. [27 pp.]. doi:10.2903/j.efsa.2010.1734. Available from: http://www.efsa.europa.eu/en/efsajournal/pub/1734.htm.

EFSA (2011), Scientific Opinion on the substantiation of health claims related to docosahexaenoic acid (DHA), eicosapentaenoic acid (EPA) and brain, eye and nerve development (ID 501, 513, 540), maintenance of normal brain function (ID 497, 501, 510, 513, 519, 521, 534, 540, 688, 1323, 1360, 4294), maintenance of normal vision (ID 508, 510, 513, 519, 529, 540, 688, 2905, 4294), maintenance of normal cardiac function (ID 510, 688, 1360), "maternal health; pregnancy and nursing" (ID 514), "to fulfil increased omega-3 fatty acids need during pregnancy" (ID 539), "skin and digestive tract epithelial cells maintenance" (ID 525), enhancement of mood (ID 536), "membranes cell structure" (ID 4295), "anti-inflammatory action" (ID 4688) and maintenance of normal blood LDL-cholesterol concentrations (ID 4719) pursuant to Article 13(1) of Regulation (EC) No 1924/2006 (question no EFSA-Q-2008-1284, EFSA-Q-2008-1288, EFSA-Q-2008-1295, EFSA-Q-2008-1297, EFSA-Q-2008-1300,…[*etc.*], adopted by written procedure on 17 February 2011 by European Food Safety Authority). *EFSA J,* **9**(4), 2078 [30 pp.].doi:10.2903/j.efsa.2011.2078. Available from: http://www.efsa.europa.eu/en/efsajournal/pub/2078.htm.

EFSA (2011), EFSA Panel on Dietetic Products, Nutrition and Allergies (NDA); Scientific Opinion on the substantiation of health claims related to glycaemic carbohydrates and maintenance of normal brain function (ID 603, 653) pursuant to Article 13(1) of Regulation (EC) No 1924/2006 (question No EFSA-Q-2008-1390, EFSA-Q-2008-1440, adopted on 08 April 2011 by European Food Safety Authority). *EFSA J*, **9**(6), 2226. [13 pp.]. doi:10.2903/j.efsa.2011.2226. Available from: http://www.efsa.europa.eu/en/efsajournal/pub/2226.htm.

IOM (2010), *Dietary reference intakes for calcium and vitamin D.* (National Academy of Sciences (NAS), Institute of Medicine (IOM), Food and Nutrition Board). Washington (DC): National Academy Press (NAP). Available from: http://www.iom.edu/~/media/Files/Report%20Files/2010/Dietary-Reference-Intakes-for-Calcium-and-Vitamin-D/Vitamin%20D%20and%20Calcium%202010%20Report%20Brief.pdf [Revised March 2011].

Regulation (EC) No 1924/2006 of the European Parliament and of the Council of 20 December 2006 on nutrition and health claims made on foods. *Off J Eur Union,* **9**(L404), 9–25. Available from http://eur-lex.europa.eu/LexUriServ/LexUriServ.do?uri=OJ:L:2006:404:0009:0025:EN:PDF.

U.S. FDA (2011), *100% Whey-protein partially hydrolyzed infant formula and reduced risk of atopic dermatitis.* Silver Spring (MD): U.S. Food and Drug Administration (U.S. FDA), Center for Food Safety and Applied Nutrition (CFSAN). Available from: http://www.fda.gov/Food/IngredientsPackagingLabeling/LabelingNutrition/ucm256731.htm [Page Last Updated: 05/26/2011].

U.S. FDA (2013a), *U.S. Code of Federal Regulations (CFR). Title 21—Food and Drugs (Food and Drug Administration).* Washington (DC): U.S. Government Printing Office (GPO). Available from: http://www.gpo.gov/fdsys/browse/collectionCfr.action?collectionCode=CFR.

U.S. FDA (2013b), *Health claims meeting significant scientific agreement (SSA).* Silver Spring, MD, U.S. Food and Drug Administration (U.S. FDA), Center for Food Safety and Applied Nutrition (CFSAN). Available from: http://www.fda.gov/Food/IngredientsPackagingLabeling/LabelingNutrition/ucm2006876.htm [Page Last Updated: 07/19/2013].

U.S. FDA (2013c), *Qualified health claims.* Silver Spring (MD): U.S. Food and Drug Administration (U.S. FDA), Center for Food Safety and Applied Nutrition

(CFSAN). Available from: http://www.fda.gov/Food/IngredientsPackagingLabeling/LabelingNutrition/ucm2006877.htm [Page Last Updated: 05/17/2013].

U.S. FDA (2013d), *Letter responding to health claim petition dated November 3, 2003 (Martek Petition): Omega-3 fatty acids and reduced risk of coronary heart disease (Docket No. 2003Q-0401)*. Silver Spring (MD): U.S. Food and Drug Administration (U.S. FDA), Center for Food Safety and Applied Nutrition (CFSAN). Available from: http://www.fda.gov/Food/IngredientsPackagingLabeling/LabelingNutrition/ucm072932.htm [Page Last Updated: 04/18/2013].

U.S. FDA (2013e), *Qualified health claims: letter of enforcement discretion – chromium picolinate and insulin resistance (Docket No. 2004Q-0144)*. Silver Spring (MD): U.S. Food and Drug Administration (U.S. FDA), Center for Food Safety and Applied Nutrition (CFSAN). Available from: http://www.fda.gov/Food/IngredientsPackagingLabeling/LabelingNutrition/ucm073017.htm [Page Last Updated: 04/18/2013].

U.S. FDA (2013f), *Qualified health claims: letters of enforcement discretion*. Silver Spring (MD): U.S. Food and Drug Administration (U.S. FDA), Center for Food Safety and Applied Nutrition (CFSAN). Available from: http://www.fda.gov/Food/IngredientsPackagingLabeling/LabelingNutrition/ucm072756.htm [Page Last Updated: 09/18/2013].

Front-of-pack (FOP) labelling of foods and beverages

7

M. Dean[1], *M. Spence*[1], *C. Hodgkins*[2], *M.M. Raats*[2]
[1]Queen's University Belfast, Belfast, UK; [2]University of Surrey, Guildford, UK

7.1 Introduction

Efforts to promote healthy food choices have gained prominence, and food packaging is seen as an ideal opportunity to encourage healthier food choice. To achieve this, nutrition labelling has been displayed on food packaging in different ways. Nutrition labelling can be viewed as a way of reducing the information asymmetry between producers and consumers, whereby producers provide consumers with information they would not otherwise have about a product (van Kleef *et al.*, 2008). This is particularly the case with complex processed foods, where recipes could vary hugely, thus making it difficult for a consumer to know the relevant proportions of risk nutrients such as fat, sugar and salt that the product may contain. However, the effectiveness of food labelling in promoting healthy food choice is hotly debated, with the motives behind industry-led schemes questioned. Some view industry-led labelling as another marketing tool, while others see it as behaviour of a threatened food sector trying to impede public health campaigns (Brownell, 2012; Brownell and Koplan, 2011).

7.1.1 What is FOP nutrition labelling?

In practice, product nutritional information has typically been communicated to consumers on a voluntary or mandatory basis, via a panel or table placed on the back or side of a pre-packaged food. Most supermarkets and many manufacturers now also display nutritional information on the front of the food package, the so-called front-of-pack (FOP) nutrition label or nutrition signpost label, which serves to provide a simple and visible summary of the product's nutritional quality. Although the use of FOP labels, logos, symbols, icons or numeric panels to communicate information is not new (developments started in the 1980s), the introduction of these systems has substantially increased over the last 10 years and the number of products displaying FOP labelling appears to have grown exponentially (Lobstein and Davies, 2008; Storcksdieck *et al.*, 2010; van Kleef and Dagevos, 2012; Williams *et al.*, 2010). A recent audit of almost 40, 000 products in all 27 EU member states and Turkey (Storcksdieck *et al.*, 2010) identified that 48% of the products displayed FOP nutrition information. Therefore, the importance of FOP nutrition labelling is increasing and has become the subject of much debate between stakeholders such as policy

Advances in Food and Beverage Labelling. http://dx.doi.org/10.1533/9781782420934.2.113

makers, health organisations, food industry and consumer associations (van Kleef and Dagevos, 2012).

The FOP nutrition label is aimed at three objectives. First, it is hoped that the FOP nutrition label will enable consumers to make faster, better informed and healthier food purchase decisions which will improve their overall diet and health (Dagevos and van Kleef, 2009; Lytton, 2010a; Williams *et al.*, 2010). Second, at an industry level, it is projected that the FOP nutrition label will motivate food manufacturers to either reformulate existing products or develop new ones, in order to provide healthier options which can carry a more favourable nutrition label (Sacks *et al.*, 2011; Vyth *et al.*, 2010). Third, it is hoped that FOP labels may be used by governments to influence public health by giving consumers informed choice while still allowing producers the freedom to manufacture what they want (van Kleef and Dagevos, 2012).

7.1.2 Types of FOP nutrition labelling

FOP labelling exists in many designs, ranging from the presence of a summary label communicating the levels of key nutrients, possibly overlaid with interpretative text or colour as a benchmark, through to the presence of a visual symbol or "health logo" indicating that the product is considered to be a more healthy choice. However, the three most prevalent labels are: the percentage guideline daily amount (GDA), the traffic-light system (TLS) and health logos (HL). GDA labels typically provide information on calories, fat, saturated fat, sugar and salt both in grams per serving and as a percentage contribution to the daily requirements of an average reference adult (Plate I in the colour section between pages 118 and 119). The TLS (Plate II) also provides information on calories and the same four key nutrients, but instead of providing GDA percentages for nutrients and the calories, these labels are colour-coded (with the exception of energy) as red, amber or green to reflect whether the nutrient content of the food is high, medium or low based on pre-established criteria (Food Standards Agency, 2007). In contrast, HLs (Plate III) do not contain numerical values for the key risk nutrients or energy content, since the presence of the logo itself indicates that the product meets the underlying nutritional criteria set by the organisation responsible for the logo scheme (Butler, 2010). In recent years, a fourth scheme, based on a hybrid (HYB) approach, has emerged, whereby both traffic lights and GDAs are displayed on the same label (Plate IV).

While a reasonable level of consensus on format and content has been achieved between policy makers on BOP nutrition labelling, with some countries opting for mandatory approaches and others voluntary, this is not yet the case for FOP nutrition labelling. The numerous FOP nutrition labels appearing on products have become a concern for governments and public health bodies, as research indicates that consumers are confused by the variety of schemes that exist (Malam *et al.*, 2009). Therefore, there has been considerable pressure on regulators to harmonise FOP labels and, as a result, many governments are discussing mandatory or voluntary guidance for placing nutrition information on the FOP (EUFIC, 2013).

7.2 Development of FOP labels

The lead for the development of a FOP label can be industry, non-governmental organisations, governments or different combinations of any of the above. In addition the nutrition composition criteria used to underpin the FOP designs can vary. These issues are discussed below.

7.2.1 History of FOP development

FOP systems initially appeared in the late 1980s and early 1990s, and were largely developed by non-profit organisations; however, as time went on industry took an active role in devising many different FOP labelling systems.

7.2.1.1 NGOs, government agencies and FOP

In 1987 the American Heart Association (AHA) created the Heart Guide symbol to signal to the consumer that a food was "heart friendly," which in 1995 was revised into the *Heart Check* programme. In the same year, the *Heart Mark* food approval programme was launched in South Africa by their Heart Foundation and, by 1992, 100 products manufactured by 18 different companies qualified to use the heart logo (a heart bracketed with a knife and fork within a circle). In 1989, the *Pick the Tick* programme was launched in Australia by the National Heart Foundation, comprising a mark of approval (a tick) in a circle surrounded by the logo of the National Heart Foundation. In 1991, a similar logo, but this time a tick within a pyramid (with a link to the healthy diet pyramid) logo, was launched in New Zealand (Graham *et al.*, 1994). Round about the same time in Canada, the Heart and Stroke Foundation created a *Health Check* programme. In contrast, in Sweden the *green keyhole symbol* was taken up by the National Food Administration (a government agency), which was then expanded to Denmark and Norway (Larsson *et al.*, 1999).

These FOP labelling programmes developed by non-governmental organisations, and the one developed by a government agency in Sweden, all utilised a symbol on the food packet to indicate its inclusion as a healthy food. The schemes differed in the selection nutrients on which the criteria were evaluated, depending on the food category (Smith *et al.*, 2002). At this time, food manufacturers were not that involved in the creation of labelling programmes or in the development of the cut-off criteria used in them.

7.2.1.2 Industry and FOP

In the UK, following collaboration between the government, consumer organisations and the food industry, the Institute for Grocery Distribution (a collective industry organisation) published guidelines for voluntary nutritional labelling including the use of GDAs for energy, fats and saturated fats (IGD, 1998). These GDAs were based on the recommendations of the Committee on Medical Aspects of Food Policy

(COMA) report on Dietary Reference values (DOH, 1991). GDAs provide guidelines on the back of pack to help consumers situate the nutrition information in the context of their overall diet. A simplified version of the GDAs as *percentage GDAs* was then used as the FOP label by many major retailers and food producers in Europe and Australia.

Also in the USA, manufacturers and retailers led the way by developing many FOP nutrition rating systems to provide consumers with easy-to-use summaries of the nutrition profile of the product. For example, Wegman's supermarkets developed a series of symbols (based on FDA and USDA nutrient content and health claims) which featured only on the store brand products, so that consumers could differentiate these products from competitors' brands. Over the years other manufacturers, such as Kraft, Unilever, Kellogg's and PepsiCo, have developed their own FOP systems. Later on algorithm-based summary symbols, such as the Hannaford Supermarkets' *Guiding Stars*, were introduced into the market place and displayed in supermarkets. This gave zero to three stars to food that met the minimum nutrient standards. The *NuVal Nutrition* scoring system is also based on an algorithm that considers and weights both positive and negative nutrients, but here the final score is presented as a number between 1and 100. Then in 2011, America's leading food and beverage manufacturers and retailers launched a new system, *Nutrition Keys*, that summarised important nutrition information (calories, saturated fat, sodium and total sugars content) on the front of food packages. In addition to the Nutrition Keys icon, some products could also display information about nutrients to encourage such as potassium, fibre, vitamin A, vitamin C, vitamin D, calcium, iron and protein. These systems were all industry-led; however, they differed in the way healthfulness was scored and displayed.

7.2.1.3 Consortiums and FOP

In addition to non-governmental organisations and industry there had been attempts for different stakeholders to get together and agree on an acceptable format for the FOPs. For example, in the USA in 2009, after a diverse group of leaders from industry, academia, public health and government had voluntarily come together for the keystone Food and Nutrition Roundtable, the *Smart Choice Programme* was developed, which started to appear on food and beverage packages. To qualify for the Smart Choice Symbol, products could not exceed standards for specific "nutrients to limit" and most categories must also provide positive attributes: to encourage nutrients or food groups. Here, criteria were set by product category (i.e. beverages, dairy, meat and snacks) rather than applying to all products. It was hoped that this would encourage smarter choices within categories, by choosing those products with icons. This category approach recognised that different foods have different profiles by nature (Lupton *et al.*, 2010).

Also in the UK after carrying out a consultation exercise on different ways of presenting nutrition information where GDAs and traffic lights were considered, the FSA proposed the TLS as the preferred scheme with the support of retailers Waitrose, Sainsbury and ASDA. Since a number of major retailers and food manufacturers (for example, Tesco and Nestlé) rejected the voluntary recommendations in favour of the GDA approach, an independent evaluation of the effectiveness of the schemes was commissioned. From these results and other consultations, the agency developed proposals to implement an HYB approach to FOP labelling that combined traffic lights and percentage GDAs.

In parallel to the on-going debate on the relative merits of GDA schemes over TLS in the UK, the European Union drafted proposals for the new legislation "Provision of food information to consumers" (EU No 1169/2011). The resulting legislation allowed for the voluntary repetition of specified nutrition information in the principle field of vision (essentially FOP), while allowing for the possibility of additional forms of expression to be used, provided they met specified criteria. That is, FOP schemes should be principally aimed at facilitating consumer understanding of the energy/nutrient contribution of the food with evidence showing that they are understood by the average consumer, and that they must not present barriers to the free movement of goods (EU No 1169/2011).

7.2.2 The underpinnings of different FOP labelling schemes

While a large variety of FOP systems have been developed and implemented by different organisations such as government, retailers, non-profit organisations, food manufacturers or a combination of these, it is worth noting that the various FOP schemes are also often underpinned by different approaches to *nutrient profiling*. Nutrient profiling is defined by the World Health Organisation (WHO) as "the science of classifying or ranking food according to their nutritional composition for reasons related to preventing disease and promoting health" (WHO, in press; Rayner *et al.*, 2013, p.1). Although this definition has been criticised for its lack of scientific rigour and the use of other food characteristics to help with the classification, Rayner *et al.* (2013) concede that despite its limitation this is the best definition available at present. In contrast, Lobstein and Davies (2008) use a more practical approach and conceptualise nutrient profiling as a "tool used to clarify the nutritional quality – the nutrient profile – of a food product both in absolute terms and in relation to other food products" (Lobstein and Davies, 2008, p.1).

7.2.2.1 Criteria- and fact-based systems

Based on the available nutrition composition data, nutrient profiling methodologies rank the healthiness of the food products using different scoring mechanisms underpinned by scientific and pragmatic principles (Townsend, 2010). Two ways of viewing the different systems is: *fact-based* and *criteria-based* systems (Pereira, 2010; Schor *et al.*, 2010). A fact-based nutrition system restates the quantitative nutrition information on the traditional BOP in a more concise fashion. For example, the GDAs display the proportion of key food components as a percentage of daily intakes of an average weight adult of average activity. The information presented could be just calories, but often includes total fat, saturated fat, sugar, sodium and sometimes fibre. GDAs can be used to get a rough idea of the ceilings for these nutrients in a typical diet (Denny,

2006). On the other hand, criteria-based systems categorise foods according to a set of criteria which can be displayed as categorical or continuous (Foltran *et al.*, 2010). In a categorical system, the labels given (e.g. one star, or three stars) are product-category specific, whereas in the continuous models (e.g. a score between 1 and 100) the same nutrient criteria can be used across different product categories. In this instance, a score is calculated for each food regardless of the product category and used to rank the foods (Foltran *et al.*, 2010). The NuVal nutritional scoring system is an example of a continuous model that uses a mathematical algorithm to compare individual products and rate them on a scale of 1–100 (Kondro, 2008). *Guiding stars* is an example of the categorical system where products with three stars are classified as healthier than those with one star.

Within the criteria-based schemes, especially in the categorical approach, different schemes vary based on the product nutrient that is emphasised. Here foods can be identified as "healthy," or "good for you" on the basis of an identified standard, such as a maximum level of fat, added sugar and sodium (Townsend, 2010). This can also lead to simple graphical symbols, such as a seal of approval, health tick, quality stamp (Butler, 2010) or labels on products, like the Swedish Green Keyhole system (Larsson *et al.*, 1999), the Smart Choices label (Lupton *et al.*, 2010) a programme used in many countries (Dötsch-Klerk and Jansen, 2008; Roodenburg *et al.*, 2009) and the PepsiCo's Smart Spot (Yach *et al.*, 2007). These and similar schemes only stress the positive aspects of the product. Conversely the Finnish "high salt" content warning is a rare example of a system that focuses on the negative aspects of a product (Pietinen *et al.*, 2008). Perhaps the only example of a system that presents both positive and negative aspects is an interpretative colour-coded scheme, such as the TLS, where thresholds for all foods are calculated using the same set of nutrients and mathematical formulae.

7.2.2.2 Qualifying and disqualifying nutrients

Nutrient profiling schemes can also differ in their choices between qualifying and disqualifying nutrients, and the choice of a reference base e.g. per 100 g/mL/kcal or serving (Drewnowski, 2007; Möser *et al.*, 2010; Verhagen and van den Berg, 2008). The various dimensions on which schemes differ are not totally independent from each other, and a combination of different approaches have been suggested, such as GDAs combined with colour-coded information on whether the percentages are favourable, neutral or unfavourable (Health Council for the Netherlands, 2008; Lobstein and Davies, 2008; Verhagen and van den Berg, 2008).

The different schemes can be compared and contrasted on different (but not completely independent) dimensions, as illustrated by Verhagen and van den Berg (2008). Here the researchers differentiated the schemes on five categories: (a) whether the scheme is based on food categories or/and in general; (b) what nutrients are used as qualifying (vitamin C, fibre, iron, protein, etc.); (c) disqualifying (salt, sugar, fat, etc.); (d) the reference base employed (100 g/100 mL, 100 kcal 100 kJ) and/or per reference quantity/serving used; and (e) whether it was underpinned by a scoring system or threshold system.

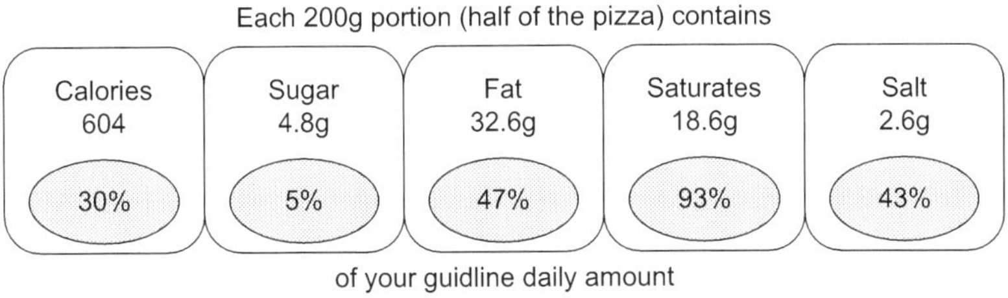

Plate I (Chapter 7) Generic example of a percentage guideline daily amount (GDA) FOP label.
Source: With permission from the University of Surrey.

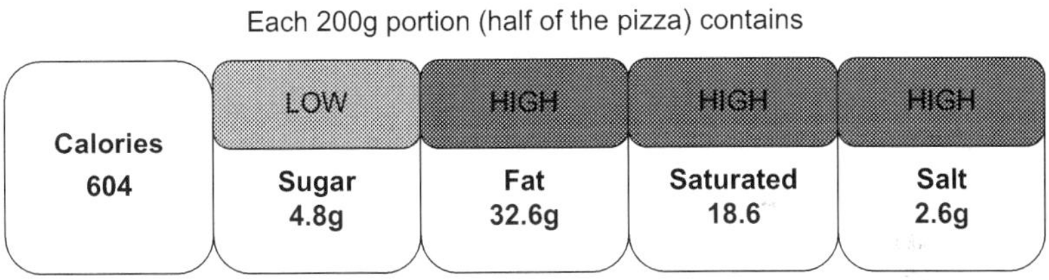

Plate II (Chapter 7) Generic example of a traffic light system (TLS) FOP label.
Source: With permission from the University of Surrey.

Plate III (Chapter 7) Generic example of a health logo (HL) FOP label.
Source: With permission from the University of Surrey.

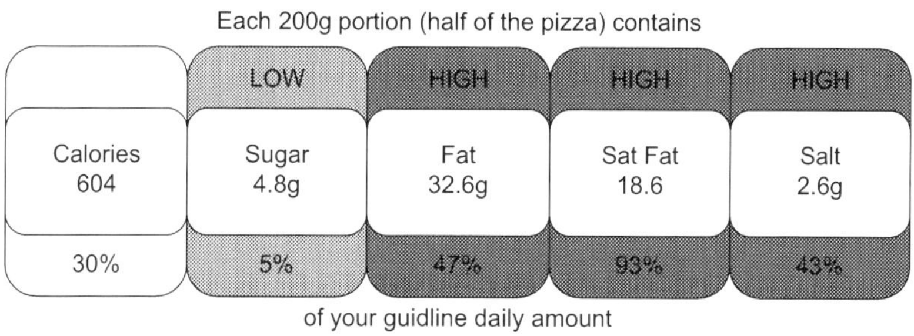

Plate IV (Chapter 7) Generic example of a hybrid system (HYB) FOP label.
Source: With permission from the University of Surrey.

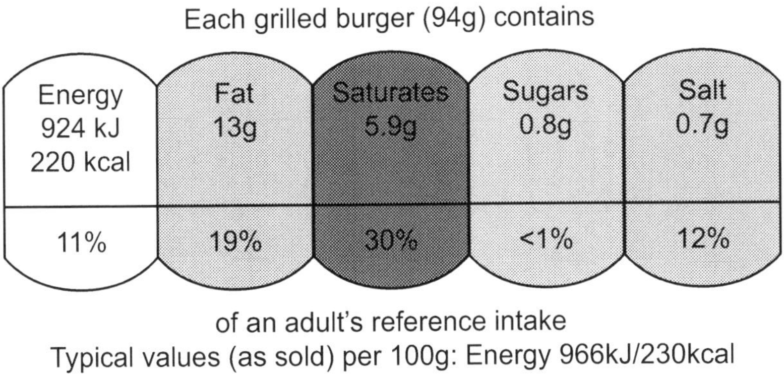

Plate V (Chapter 7) Recommended UK Government FOP label.
Source: Public sector information licensed under the Open Government Licence v2.0, http://www.nationalarchives.gov.uk/doc/open-government-licence/version/2/'.

Plate VI (Chapter 11) Thermochromic label on Coors Light beer bottle, illustrating label before and after chilling.
Source: Image courtesy of LCR Hallcrest.

Plate VII (Chapter 11) Fresh-Check® indicator illustrating the progressive darkening of central circle.
Source: Image courtesy of TempTime Corporation.

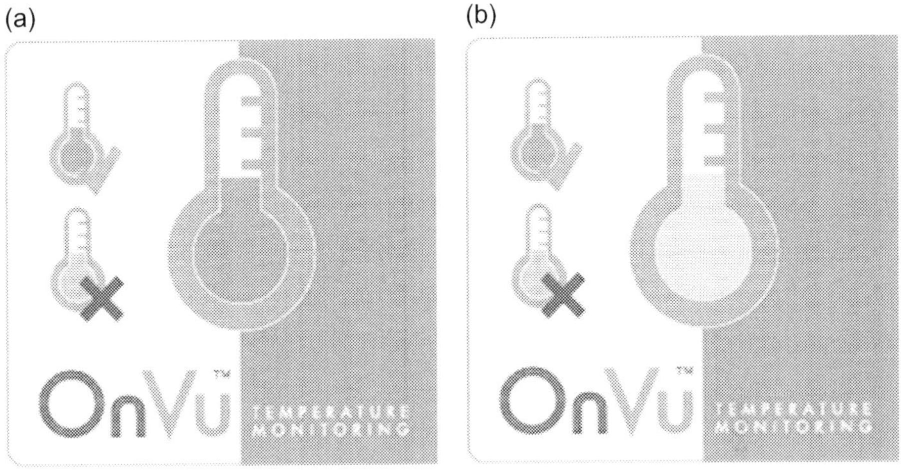

Plate VIII (Chapter 11) OnVu™ label illustrating the dark blue (a) and light blue (b) stages of the colour change (image courtesy of Bizerba, Marc Büttgenbach (marc.buettgenbach@bizerba.com) and Pedro Garza (pedro.garza@bizerba.com)).

Plate IX (Chapter 11) Image of sensor foil which changes colour due to high amine concentration in the foodstuff.
Source: Image courtesy of © Fraunhofer EMFT.

Plate X (Chapter 11) Image of a RipeSense™ label.
Source: Image courtesy of Jenkins Freshpac Systems, www.jenkinsfps.co.nz.

Plate XI (Chapter 11) Insignia Technologies Ltd.'s 'embedded timer' technology.
Source: Image courtesy of Insignia Technologies Ltd.

Plate XII (Chapter 11) Insignia Technologies Ltd.'s CO_2 fruit transportation label.
Source: Image courtesy of Insignia Technologies Ltd.

7.2.3 Classifications of FOP schemes

In addition to the classification system used described in the previous section, over recent years other categorisation schemes or "taxonomies" that encompass the various FOP schemes have evolved. The ability to evaluate the widely differing FOP schemes across a number of relevant dimensions is important when policy makers consider their potential merits and possible areas where standardisation or control may be required in the form of legislation.

After reviewing 20 different systems that were representative of those available in the US market place, the IOM (Institute of Medicine) committee categorised these systems as nutrient specific systems, summary indicator systems and food group information systems (IOM, 2010). Each system used different criteria to rate and compare foods, primarily within a category. In contrast, Lytton (2010b) proposed a taxonomy which compares and contrasts the schemes across a number of dimensions including "source, scope, character, gradation, segmentation and aggregation" (Lytton, 2010b). While the above taxonomies are useful to compare the differing FOP labelling approaches from an expert perspective, Pereira (2010) argues that consumers are unlikely to recognise the differences between the various dimensions and there is a need to understand how consumers view the different schemes on the market.

In addressing this need, Hodgkins *et al.* (2012) developed a taxonomy based on "directiveness" as a means of differentiating between the FOP schemes from a consumer perspective. On this dimension, three categories are proposed: *Directive, Non-directive and Semi-directive.* FOP labels that communicate, typically via a logo or symbol, whether a product is considered healthy in relation to other products in the same food category are classified as *Directive* by this taxonomy. The presence of the logo communicates that a decision as to the healthfulness of the product has already been made on behalf of the consumer and therefore these labels do not display any detailed nutrition information for the consumer to process (see Plate III). In contrast, FOP labels containing detailed numerical information (typically on energy, fat, saturated fat, salt and sugar per serving), such as the GDA scheme, do not communicate an overall decision on the healthfulness of the product, or give any benchmarks about the nutrient levels in the food. With these labels, the consumers have to decide for themselves how healthy the product is by using the nutritional information provided. Therefore, these schemes are classified as *Non-directive* (see Plate I). Finally, if any detailed numerical information in the label is overlaid with colour or text as an additional form of interpretative information indicating whether the nutrient levels in the food are healthful or not, the labelling scheme is classified as being *Semi-Directive*. While the semi-directive FOP labels, such as the TLS or HYB schemes, do not formally communicate an overall decision on whether the product as a whole is considered to be healthy, they are more in line with the directive schemes when the colour coding is predominantly red or green across the nutrients. However, when a mixture of red, green and amber is present across the nutrients on a single food, they become non-directive at a product level but still retain some directiveness at a nutrient level (see Plates II and IV).

While the expert- as well as consumer-led classification schemes may be critiqued, the value of a particular classification scheme depends on what is hoped to be achieved. Ultimately, what is important is whether consumers know what is being communicated to them through labelling, and whether they know what they are meant to do with the FOP labels provided. Further, there is a need to understand which types of decisions can be usefully supported by FOP labels. For example, is labelling best for making choices within a food category or to make choices between categories? Also, we need to know more about how consumers use FOP labels to make decision on new products.

7.3 Impact of FOP labelling

The impact of FOP labelling can be evaluated based on the objectives FOP labelling was intended to achieve. Thus, the direct impact of FOP can be assessed in relation to the consumer as well as the food supply (Smith *et al.*, 2002). In addition, the direct impact of FOP on consumer behaviour can be measured by looking at the healthfulness of the food purchased and consumers' dietary intakes. The indirect impact of FOP on the consumer can be explored by various measures such as enhanced consumer knowledge, the use of FOP in decision making, interpretation of FOP and increased consumer awareness of FOP. The impact of FOP on the food supply can be measured by the level of adoption of FOP labelling, as well as the amount of new products developed and/or reformulated that are available on the market displaying more healthful FOP profiles.

7.3.1 Consumers

Grunert and Wills (2007) stress that FOP labels should be "easy to use, carry relevant nutritional information and be non-coercive" (Grunert and Wills, 2007). However, emerging evidence suggests that the multitude of schemes and their differing presentations on packs are confusing the consumers, rather than helping them to make healthier decisions (Malam *et al.*, 2009).

7.3.1.1 FOP labels and healthy food choice

There is little evidence that consumers are able to effectively utilise GDAs in the context of their overall daily diet; however, research suggests that consumers are able to identify more healthful products by using GDAs (Grunert *et al.*, 2010). Further, the inherent flexibility of the GDA scheme allowing presentation of GDA values in different portion sizes has been criticised. Differences in portion sizes can make it difficult for consumers to compare products (Lobstein and Davies, 2008), and can even make the product appear more healthful if unrealistic portion sizes are used as the basis for the nutritional information in the FOP label. In contrast, supporters of HL schemes argue that by pre-assessing both the positive and negative nutrients, these schemes communicate the healthfulness of the product as a whole, without the need for any numerical processing by the consumer at point-of-purchase. However, Feunekes *et al.* (2008) found that consumers preferred the TLS to a HL scheme for liking, comprehension and credibility. Between HL schemes, concerns have also been raised regarding the different thresholds applied, so that not all foods displaying logos are essentially healthful foods, but may just be the most healthful profiled within a particular food category. While there is growing evidence to suggest that the TLS and other colour-coded FOP labelling schemes may be the most promising at encouraging healthful food choices (Lobstein and Davies, 2008; Thorndike *et al.*, 2012), there is some concern, particularly within industry, that the red colour may be interpreted to mean "not allowed" rather than "limit intake". This may result in avoidance of important food groups which are essential for a well-balanced diet (Grunert and Wills, 2007).

Van Herpen *et al.* (2012), investigating the effect familiarity has on FOP evaluation and use in the UK and the Netherlands using a laboratory experimental design, found that all labels were equally effective in enhancing healthful choice. They claim that although non-directive labels (such as nutrition labels and mono-chrome GDA) are evaluated more positively, other labels (TLS, Choices logo) are equally, or even more, effective in enhancing the healthful choices. Van Herpen *et al.* (2012) argue that self-reported evaluation and usage intentions are not accurate predictors of FOP's effects on healthful product choices. They found that both in the UK, where TLS and GDA labelling prevail, and in the Netherlands, where the Choices logo and GDA are most widely used, all three labels were effective in enhancing healthful product choices despite prior familiarity with the labelling schemes affecting consumers' self-reported evaluation and usage intentions. They conclude that in terms of healthfulness perceptions, all labels reward the more healthful options within the assortment but only TLS and GDA also disqualify other options, especially in the medium healthful product options (Van Herpen *et al.*, 2012).

Contrasting different types of FOP labels in experimental studies has shown that the majority of people understand or can successfully identify more healthful products using any of the prominent labelling formats (FLABEL, 2012; Watson, 2014). However, in an experimental study to explore health motivation and perceived capability of making healthful choices using a variety of FOP labels, colour coding was found to improve the healthfulness of choices when participants were motivated to choose a healthful product, but not when they were asked to choose according to their preferences (Aschemann-Witzel *et al.*, 2013). In addition, the study demonstrated that manipulating the choice environment by increasing the availability of healthier options may in fact be more effective. Therefore, it can be argued that just because consumers armed with some basic nutritional information can differentiate between alternative foods under these experimental conditions, this does not necessarily mean that they can do the same in the real world, where other factors such as motivation and time to process nutritional information play a big part (Gerrior, 2010), as does the choice environment itself. Thus, so far there is no strong supporting evidence to suggest that providing FOP labels is having the desired effect on consumers making healthful choices in real world situations. This seems to be the case even when FOP label presentation has been standardised, increasing consistency and consumer attention to FOP labelling. That is, there is no evidence to suggest consistent labelling will translate into more healthful purchasing (Koenigstorfer *et al.*, 2013; Van Herpen and Van Trjip, 2011).

7.3.1.2 FOP labelling and healthy food sales in different settings

Cafeteria sales data (Vyth *et al.*, 2011) and change in sales before and after the introduction of GDA labels (Boztug *et al.*, 2012) all show no significant increase in healthful food purchase. While there is evidence that TLSs are favoured by consumers (Hawley *et al.*, 2012; Lachat and Tseng, 2013), the introduction of TLS labelling was not found to have an impact on the healthfulness of food purchases in the first 4 weeks of the TLS introduction in a UK retailer (Sacks *et al.*, 2009). The limited range of food categories studied, as well as the short time frame, may account for some of the findings.

In contrast, a significant change in healthier purchasing was observed in a hospital intervention study after the introduction of the TLS (Thorndike *et al.*, 2012). However, in this study, the traffic-light labelling was a single colour indicator at product level, and additional support was given, in the form of signage at point of sale and the presence of a dietician to assist with consumer choice queries. A further increase in healthier foods was observed when healthier products (green labelled) were placed at eye level at the point of sale. This suggests that labelling alone may not be enough and additional supports in terms of placement/trouble shooting are needed for healthful purchasing.

Hersey *et al.* (2013) reviewed the effects of FOP and shelf nutrition labelling systems on consumers and identified studies relating to purchase behaviour and to consumption (likely, reported and observed consumption). The results relating to purchase behaviour showed that in four out of the six empirical studies consumers were influenced to buy products indicated as healthy by shelf-labelling systems, and the studies that looked at consumers' purchase behaviour after the introduction of multiple level summary icon shelf tags in supermarkets found consumers to be influenced to buy healthier products as a result of the icons. For example the 'Guiding Stars' programme used in one of the supermarkets in the USA saw increases in sales of starred frozen dinners and yoghurts by 56% and 8% respectively, and outsold un-starred dinners and yoghurt in the first year. Also, sales in lean ground beef increased by 18%, whereas fattier ground meat decreased by 56%. Another study reported a 0.5% increase in starred foods after 1 year and a 1.4% increase in sales of starred foods after 2 years across 168 supermarkets located in the north-eastern United States. In contrast, a study assessing the effects of introducing "Fuel Your Life" shelf tags in one US college convenience store found no significant effect on sales of products tagged as nutritious (Freedman and Connors, 2010).

7.3.1.3 FOP and consumer diets

The influence of FOPs and shelf-labelling tags on consumer selection of healthy foods and healthy diets studies is limited and the results mixed (Hersey *et al.*, 2013).

Hersey *et al.*, who evaluated the effect of the Choices logo in a Dutch population, found no significant differences between the logo and no-logo conditions in participants' actual consumption or on other measures. Another simulated shopping study, conducted in Germany, examined consumers' food and drink selections on what they intended to consume the following day, and found no differences among the five FOP label schemes: a single level summary icon; grams, TLS, text; %GDA, grams; %GDA, grams, TLS; and no label (Borgmeier and Westenhoefer, 2009). A Swedish study found consumers (with the exception of the least educated) who had knowledge of the Swedish Keyhole symbol (a single summary icon) reported consuming more low-fat foods than consumers who did not have knowledge of this symbol (Larsson *et al.*, 1999). A Canadian study found that consumers who purchased foods with the "Health Check" logo (again a single summary icon) reported that they had lower fat diets than those who did not purchase products with the logo (Reid *et al.*, 2004).

Many of the studies were conducted in laboratories and very few investigated consumer behaviour in a shopping environment. Furthermore, in these studies, some research looked at actual purchase of foods and the use of FOP labels, while other relied on self-reported purchase or intention to purchase. However, as only one study measured consumers' actual intake before and after FOP implementation there is limited evidence that FOP and shelf tags actually change consumers' eating behaviour.

In summary, there does not seem to be any consensus as to whether FOPs do have a positive impact on consumers' diets. Results are difficult to compare, since the outcome variables (liking, intention, reported behaviour, actual behaviour) differ across the studies. Also, depending on where the study was conducted i.e. lab, cafe, shop or hospital, differing results are often obtained. In addition, whether or not consumers were given additional help while using the FOP label, as well as the length of period of observation and the food categories included in the study, all appear to confound the findings.

7.3.2 Industry

The impact of FOP to the food supply can be investigated by looking at food manufacturers' responses in relation to the number and types of new products developed as well as the level of reformulation of already available products (Golan *et al.*, 2009).

7.3.2.1 Adoption of FOP on food products

Using the Global New Product Database (GNPD), Mintel, Van Camp and Hooker (2010) analysed the number of innovative and reformulated packaged food products released in the UK between 2002 and 2008, dividing the labelled foods into a *target group* – those that included FSA signposting campaign foods (hot and cold cereals, pastry dishes, pizzas, prepared meals and sandwiches) and a *general group* (bread, cakes/pies/sweet goods, savoury biscuits, sweet biscuits, instant noodles/pasta/rice and meal kits).

In terms of adoption of FOP labelling, Van Camp and Hooker (2010) found there was an increased use of TLS and GDA labels and a decline in the use of other labels after 2005 in both groups. GDA had a higher percentage of adoption than TLS in both groups, but both GDA and TLS were adopted a year earlier and on a larger share of foods in the target group (hot and cold cereals, pastry dishes, pizzas, prepared meals and sandwiches) than in the general group (bread, cakes/pies/sweet goods, savoury biscuits, sweet biscuits, instant noodles/pasta/rice and meal kits). In 2008, 55% of all products released in the target group had some type of FOP nutrition labelling, compared to only 27% of all products in the general group. The share of GDA and TLS labelled foods in the target group were 42% and 26%, while the share for the general group was only 20% and 4% (Van Camp and Hooker, 2010).

They also found that FOP adoption strategies varied significantly between private-label and branded products. None of the leading branded food manufacturers had adopted the TLS labelling, and even among retailers the use of TLS labels had been selective across food categories. For example, although the UK retailers Sainsburys and Waitrose had committed to use only TLS labelling, they did not use FOP labelling on all products. In 2008, Sainsburys only had 5.4% of all innovations in the general group and only used TLS labelling on 3.6% of products. On the other hand, Tesco and Somerfield, like the leading brands, had chosen to use only GDA labelling, but again did not use FOP labelling on all products. Marks and Spencers (M&S) and ASDA had opted to use a combination of TLS and GDA label, and both used the label much more frequently on foods that were in the target categories. For example, in 2008, M&S and ASDA labelled 80% and 52% of food products released in the target group respectively, whereas the combined label was used only on one product in the general group.

Thus, research shows that the voluntary adoption of TLS and GDA among food manufacturers has been widespread across different food categories, with GDA the most widely adopted labelling scheme both in the private and branded firms in the UK. The adoption of GDA rather than TLS labelling may be attributed to the influence of proposed EU legislation and competitive pressure for manufacturers to reformulate products. Food manufacturers may have thought it better to align with the EU regulators rather than the national schemes in order to reduce labelling costs and consumer confusion.

7.3.2.2 Reformulation of products

Vyth *et al.* (2012) identified three published studies that had evaluated the effects of FOP on product development. In one study (Vyth *et al.*, 2010), 47 manufacturers (response rate of less than 40%) who had joined the Choices foundation in Netherlands indicated that out of 821 products (23.5% of all choice products available) 417 were on products that were on already existing products, 168 had been reformulated and 236 were newly developed. From the data collected, most products carrying the Choices logo as a result of reformulation were on soups, followed by sandwiches. The largest number of products carrying the logo as a result of new product development was snacks, followed by processed fruits and vegetables, fruit juices and soups. Sodium reduction was the most common change found in processed meats, sandwiches, soups and sandwich fillings. Dietary fibre was significantly increased in most newly developed product groups, for example, in fruit juices, processed meats, dairy products, sandwiches and soups. Saturated fatty acids and added sugar were significantly decreased both in reformulated and newly developed dairy products. However, calorie content was significantly reduced only in reformulated dairy products, sandwich fillings and in some newly developed snacks. Vyth *et al.* (2010) argue that the Choices logo has motivated food manufacturers to reformulate existing products and develop new products with healthier composition in the sodium and dietary fibre areas, although they do note that the data gathered were self-reported and not observed in person.

Williams *et al.* (2003) studied salt reduction in Kellogg's breakfast cereals, and they also found that by using the Pick the Tick criterion for salt as the target value, 12 cereal products were reformulated with an average reduction of 40%. Williams argues that the impact of salt reduction was not only on the foods that were part of the food approval programme, but also extended to other foods in the food supply chain. Similar findings were reported by Young and Swinburn (2001) in New Zealand in relation to the Pick the Tick programme there.

In sum, there is evidence to show that independent companies are manufacturing innovative products in an attempt to display more favourable FOP profile labels through reformulation and new product development. However, more evidence from a wider variety of countries is needed before any strong conclusions may be reached.

7.4 Future trends and developments

From a consumer perspective it could be argued that there is little supporting evidence that FOP labels are having the desired effect on dietary choices. Thus there is an urgent need to determine how and why consumers are using FOP labels and to investigate ways of motivating consumers to use FOP labels. Further work is needed to better understand the role of labelling in the broader set of policy tools available to influence food production and consumption. The way and extent to which FOP labelling impacts on reformulation of products also warrants further study.

7.4.1 On-going regulatory discussion

The European Food Information Regulation (EU No 1169/2011) mandatory nutrition labelling component comes into force in 2016. Under the regulation, FOP labelling is not compulsory, but in 2013 the UK Government announced their final recommendation for an FOP nutrition labelling which will involve traffic-light colours and percentage GDAs in a HYB format (DOH, 2013a, b). This FOP label (Plate V) has been agreed by all the major retailers and supermarkets in the UK, as well as Nestlé, Mars and Pepsi, and those companies that have signed up to using the label are said to account for more than 60% of the food sold in the UK (DOH, 2013a). However, multinationals such as Kraft, Unilever, Kellogg's and Coca-Cola are not yet supportive of this scheme. Further, the scheme is not compulsory as per EU law, and this position is not envisaged to change anytime in the near future.

Despite this emerging consensus on a UK FOP label format, there is still on-going debate as to which nutrients should be highlighted in the label, with some nutritionists advocating energy (calories), saturated fat, salt and dietary fibre, and not fat and sugar (Rayner, 2013). There is also strong opposition from other EU countries to the UK approach.

7.4.2 Enhancing consumer and industry impact

In relation to consumer impact, research shows that simply providing nutrition information on the BOP and highlighting it on the FOP is not effectively resulting in more healthful food purchasing behaviour, as in real life settings, personal and contextual factors interfere and some of these take precedence over health consideration in driving choice (Barker *et al.*, 2012; FLABEL, 2012). Further, changing behaviour can be difficult and often requires intensive interventions that are multi-faceted. Thus, future research into FOP labelling should focus on the development of a greater understanding of the psychosocial and contextual factors impacting on the performance of an FOP system in a real world environment. The efficacy of FOP labels to change consumers' eating behaviour, and thereby combat obesity, should be investigated so that realistic achievable targets can be set and success assessed accordingly.

Further there should be more lucidity on what, when and how labels are expected to be used by consumers. For example, should consumers make radical shifts to the diets they consume, or should they make smaller shifts within category? At a societal level, there should be clarity about how and where the food supply chain is expected to change; should there be less production of unhealthy foods or should there be a reduction of the portion sizes of less healthy foods? Furthermore, as argued by Gortmaker *et al.* (2013), a more comprehensive monitoring system is needed to establish the impact labelling is having on the supply chain in terms of reformulation and new product development.

7.5 Sources of further information

Campos, S., Doxey, J. and Hammond, D. (2011) "Nutrition labels on pre-packaged foods: a systematic review," *Public Health Nutr*, **14**(8), 1496–1506.

Cowburn, G. and Stockley, L. (2005) "Consumer understanding and use of nutrition labelling: a systematic review," *Public Health Nutr*, **8**(1), 21–28.

Foltran, F., Verduci, E., Ghidina, M., Campoy, C., Jany, K-D., Widhalm, K., Biasucci, G., Vögele, C., Halpern, G.M. and Gregori, D. (2010) "Nutritional profiles in a public health perspective: a critical review," *J Int Med Res*, **38**, 318–385.

Grunert, K.G. and Wills, J.M. (2007) "A review of European research on consumer response to nutrition information on food labels," *J Public Health*, **15**, 385–399.

Hawley, K.L., Roberto, C.A., Bragg, M.A., Liu, P.J., Schwartz, M.B. and Brownell, K.D. (2012) "The science on front-of-package food labels," *Public Health Nutr*, 1–10.

Hersey, J.C., Wohlgenant, K.C., Arsenault, J.E., Kosa, K.M. and Muth, M.K. (2013) "Effects of front-of-package and shelf nutrition labeling systems on consumers," *Nutr Rev*, **71**(1), 1–14.

Pereira, R. (2010) "Understanding the maze of nutrition symbols and nutrition labeling systems" *Diabetes Spectr*, **23**(1), 57–63.

Acknowledgements

The authors would like to acknowledge Nicky Charles (University of Surrey) for her assistance in preparing this chapter.

References

Aschemann-Witzel, J., Grunert, K.G., Van Trijp, H.C.M., Bialkova, S., Raats, M.M., Hodgkins, C., Wasowicz -Kirylo, G. and Koenigstorfer J. (2013) "Effects of nutrition label format and product assortment on the healthfulness of food choice," *Appetite*, **71**, 63–74.

Barker, M., Lawrence, W. and Robinson S. (2012) "Food labelling and dietary behaviour: bridging the gap," *Public Health Nutr*, **15**(5), 758–759.

Borgmeier, I. and Westenhoefer, J. (2009) "Impact of different food label formats on healthiness evaluation and food choice of consumers: a randomized-controlled study," *BMC Public Health*, **9**, 184.

Boztuğ, Y., Juh, H.J., Elshiewy, O. and Jensen, M.B. (2012) "Are consumers influenced in their food choice by health labels?" In *Proceedings of the 41th EMAC Conference*, 22–25 May 2012; Lisbon, Portugal. [Online]. Available at: http://www.emac2012.org/userfiles/emac2012_conferenceproceedings_final%20may%2017.pdf (Accessed: 2 June 2014).

Brownell, K.D. (2012) "Thinking forward: the quicksand of appeasing the food industry," *PLoS Med*, **9**(7), e1001254, doi:10.1371/journal.pmed.1001254.

Brownell, K.D. and Koplan, J.P. (2011) "Front- of-pack nutrition labeling – an abuse of trust by the food industry," *N Engl J Med*, **364**(25), 2373–2375.

Butler, K.M.L. (2010) "*Making Smart Choices: Health Claims, Regulation, and Food Packaging.*" PhD thesis. University of Pittsburgh, Pittsburgh. [Online]. Available at: http://d-scholarship.pitt.edu/7067/1/Butler_K_April_2010.pdf (Accessed: 2 June 2014).

Dagevos, H. and van Kleef, E. (2009) *Gezondheidslogo's op eten. Verkenningen rond hun recente opmars*. Wageningen: Wageningen Academic Publishers. 133–142. ISBN 9789086861361.

Denny, A. (2006) "Stop, think, go? Are signposting labelling schemes the way forward?," *Nutr Bull*, **31**(2), 84–87.

Department of Health, UK (DOH) (1991) *Dietary Reference Values for Food Energy and Nutrients for the United Kingdom*. no. 41. London: HMSO. [Online]. Available at: http://www.sacn.gov.uk/pdfs/sacn_02_26.pdf (Accessed: 2 June 2014).

Department of Health, UK (DOH). (2013a) *Final Design of Consistent Nutritional Labelling System Give Green Light [Press Release] 19 June 2013*. [Online]. Available at: https://www.gov.uk/government/news/final-design-of-consistent-nutritional-labelling-system-given-green-light (Accessed: 2 June 2014).

Department of Health, UK (DOH). (2013b) *Guide to Creating a Front of Pack (FOP) Nutrition Label for Pre-Packed Products Sold Through Retail Outlets Guidance Document.* [Online]. Available at: https://www.gov.uk/government/uploads/system/uploads/attachment_data/file/300886/2902158_FoP_Nutrition_2014.pdf (Accessed: 2 June 2014)

Dötsch-Klerk, M. and Jansen, L. (2008) "The Choices programme: a simple, front-of-pack stamp making healthy choices easy," *Asia Pac J Clin Nutr*, **17** (S1), 383–386.

Drewnowski, A. (2007) "What's next for nutrition labeling and health claims? An Update on Nutrient Profiling in the European Union and the United States," *Nutr Today*, **42**(5), 206–214.

EU No 1169/2011 Regulation of the European Parliament and of the Council of 25 October 2011 on the provision of food information to consumers, [2011] OJ L304/18. [http://eurlex.europa.eu/LexUriServ/LexUriServ.do?uri=OJ:L:2011:304:0018:0063:EN:PDF.] (Accessed: 2 June 2014)

European Food Information Council (EUFIC). (2013) *Global Update on Nutrition Labelling*. Executive summary. [Online]. Available at: http://www.eufic.org/upl/1/default/doc/GlobalUpdateExecSumFeb2013_FINAL.pdf (Accessed: 2 June 2014).

Feunekes, G.I.J., Gortemaker, I.A., Willems, A.A., Lion, R. and Van den Kommer, M. (2008) "Front-of-pack nutrition labelling: testing effectiveness of different nutrition labelling formats front-of-pack in four European countries," *Appetite*, **50**(1), 57–70.

FLABEL consortium. (2012) Food Labelling to Advance Better Education for Life – Webinar announcing final results. [Online]. Available at: http://flabel.org/en/News/FLABEL-final-webinar/. (Accessed: 2 June 2014).

Foltran, F., Verduci, E., Ghidina, M., Campoy, C., Jany, K-D., Widhalm, K., Biasucci, G., Vögele, C., Halpern, G.M. and Gregori, D. (2010) "Nutritional profiles in a public health perspective: a critical review," *J Int Med Res*, **38**, 318–385.

Food Standards Agency (2007) *Front of Pack Traffic Light Technical Guidance*, Issue 2. London: Food Standards Agency. [Online]. Available at: http://www.food.gov.uk/multimedia/pdfs/frontofpackguidance2.pdf. (Accessed: 2 June 2014).

Freedman, M.R. and Connors, R. (2010) "Point-of-purchase nutrition information influences food-purchasing behaviors of college students: a pilot study," *J Am Diet Assoc*, **110**, 1222–1226.

Gerrior, S.A. (2010) "Nutrient profiling systems: are science and the consumer connected," *Am J Clin Nutr*, **91**(4), 1116S–1117S.

Golan, E., Mancino, L. and Unnevehr, L. (2009) "Food policy: check the list of ingredients," *Amber Waves,* **7**(2), 16–21.

Gortmaker, S.L., Story, M., Powell, L.M. and Krebs-Smith, S.M. (2013) "Building infrastructure to document the US food stream," *Am J Prev Med*, **44**(2), 192–193.

Graham, P., Harker, D., Harker, M. and Tuck, M. (1994) "Branding Food Endorsement Programs. The National Heart Foundation of Australia," *J Prod Brand Manag*, **3**(4), 31–43.

Grunert, K.G. and Wills, J.M. (2007) "A review of European research on consumer response to nutrition information on food labels," *J Public Health*, **15**, 385–399.

Grunert, K.G., Fernández-Celemín, L., Wills, J.C., Storcksdieck genannt Bonsmann, S. and Nureeva, L. (2010) "Use and understanding of nutrition information on food labels in six European countries," *J Public Health*, **18**, 261–277.

Hawley, K.L., Roberto, C.A., Bragg, M.A., Liu, P.J., Schwartz, M.B. and Brownell, K.D. (2012) "The science on frontofpackage food labels," *Public Health Nutr*, **16**(3), 430–439.

Health Council of the Netherlands (2008) *Healthy Nutrition: A Closer Look at Logos.* [Online]. Available at: http://www.gezondheidsraad.nl/sites/default/files/200822er_0.pdf (Accessed: 2 June 2014).

Hersey, J.C., Wohlgenant, K.C., Arsenault, J.E., Kosa, K.M. and Muth, M.K. (2013) "Effects of front-of-package and shelf nutrition labeling systems on consumers," *Nutr Rev*, **71**(1), 1–14.

Hodgkins, C., Barnett, J., Wasowicz-Kirylo, G., Stysko-Kunkowska, M., Gulcan, Y., Kustapeli, Y., Akgungor, S,. Chryssochoidis, G., Fernández-Celemín, L., Storcksdieck genannt Bonsmann, S., Gibbs, M. and Raats M. (2012) "Understanding how consumers categorise nutritional labels: a consumer derived typology for front-of-pack nutrition labelling," *Appetite*, **59**, 806–817.

Institute of Grocery Distribution (IGD) (1998) *Voluntary Nutrition Labelling Guidelines to Benefit the Consumer* (01.02): IGD.

IOM (Institute of Medicine). (2010) *Examination of Front-of Package Nutrition Rating Systems and Symbols: Phase I Report.* Washington, DC.: The National Academies Press.

Koenigstorfer, J., Wąsowicz-Kiryło, G., Styśko-Kunkowska, M. and Groeppel-Klein, A. (2013) "Behavioural effects of directive cues on front-of-package nutrition information: the combination matters!" *Public Health Nutr*,**16**, 1–7.

Kondro, W. (2008) "Canada needs paradigm shift in public health nutrition," *Can Med Assoc J*, **179**(12), 1259–1261.

Lachat, C. and Tseng, M. (2013) "A wake-up call for nutrition labelling," *Public Health Nutr*, **16**(3), 381–382.

Larsson, I., Lissner, L. and Wilhelmsen L. (1999) "The 'Green Keyhole' revisited: nutritional knowledge may influence food selection," *Eur J Clin Nutr*, **52**, 776–780.

Lobstein, L. and Davies, S. (2008) "Defining and labelling 'healthy' and 'unhealthy' food," *Public Health Nutr*, **12**(3), 331–340.

Lupton, J.R., Balentine, D.A., Black, R.M., Hildine, R., Ivens, B.J., Kennedy, E.T., Packard, P. T., Sperber, B.R., Steffen, D. and Story, M. (2010) "The smart choices front-of-package nutrition labeling program: rationale and development of the nutrition criteria," *Am J Clin Nutr*, **91**(4), 1078S–1089S.

Lytton, T.D. (2010a) "Banning front-of-package food labels: First Amendment constraints on public health policy," *Public Health Nutr*, **14**(6), 1123–1126.

Lytton, T.D. (2010b) "Signs of change or Clash of Symbols? FDA Regulation of nutrient profile labelling," *Health Matrix*, **20**, 93–144.

Malam, S., Clegg, S., Kirwan, S. and McGinigal, S., in association with Raats, M., Barnett, J., Senior, V., Hodgkins, C. and Dean, M. (2009) *Comprehension and Use of UK Nutrition Signpost Labelling Schemes.* Report prepared for the Food Standards Agency. London: British Market Research Bureau with University of Surrey. [Online]. Available at: http://multimedia.food.gov.uk/multimedia/pdfs/pmpreport.pdf (Accessed: 2 June 2014).

Möser, A., Hoefkens, H., Van Camp, J. and Verbeke, W. (2010) "Simplified nutrient labelling: consumers' perceptions in Germany and Belgium," *Journal für Verbraucherschutz und Lebensmittelsicherheit*, **5**(2), 169–180.

Pereira, R. (2010) "Understanding the maze of nutrition symbols and nutrition labeling systems," *Diabetes Spectr*, **23**(1), 57–63.

Pietinen, P., Valsta, L.M., Hirvonen, T. and Sinkko, H. (2008) "Labelling the salt content in foods: a useful tool in reducing sodium intake in Finland," *Public Health Nutr*, **11**(4), 335–340.

Rayner, M. (2013) *Food Traffic Lights are Green for Go but EU Holds Back More Radical Measures.* The conversation. The Conversation Media Group. [Online]. Available at: http://theconversation.com/food-traffic-lights-are-green-for-go-but-eu-holds-back-more-radical-measures-15403 (Accessed: 2 June 2014).

Rayner, M., Scarborough, P. and Kaur, A. (2013) "Nutrient profiling and the regulation of marketing to children. Possibilities and pitfalls," *Appetite*, **62**, 232–235.

Reid, R.D., Slovenic D'Angelo, M.E., Dombrow, C.A., Heshka, J.T. and Dean, T.R. (2004) "The Heart and Stroke Foundation of Canada's Health Check Food Information Program," *Can J Public Health*, **95**(2), 146–50.

Roodenburg, A.J.C., Temme, E.H.M., Davies, O.H. and Seidell, J.C. (2009) "Potential impact of the Choices Programme on nutrient intakes in the Dutch population," *Nutr Bull*, **34**, 318–323.

Sacks, G., Rayner, M. and Swinburn, B. (2009) "Impact of front-of-pack 'traffic-light' nutrition labelling on consumer food purchases in the UK," *Health Promot Int*, **24**(4), 344–352.

Sacks, G., Veerman, J.L., Moodie, M. and Swinburn, B. (2011) "Traffic-light' nutrition labelling and 'junk-food' tax: a modelled comparison of cost-effectiveness for obesity prevention," *Int J Obes*, **35**(7), 1001–1009.

Schor, D., Maniscalco, S., Tuttle, M.M., Alligood, S. and Reinhardt Kapsak W. (2010) "Nutrition facts you can't miss: the evolution of front-of-pack labeling. providing consumers with tools to help select foods and beverages to encourage more healthful diets," *Nutr Today*, **45**(1), 22–32.

Smith, S.C., Stephen, A.M., Dombrow, C. and MacQuarrie, D. (2002) "Food information programs: a review of the literature," *Can J Diet Pract Res*, **63**(2), 55–60.

Storcksdieck genannt Bonsmann, S., Fernández-Celemín, L., Larrañaga, A., Egger, S., Wills, J.M., Hodgkins, C. and Raats, M.M. (2010) "Penetration of nutrition information on food labels across the EU-27 plus Turkey," *Eur J Clin Nutr*, **64**, 1379–1385.

Thorndike, A.N., Sonnenberg, L., Riis, J., Barraclough S. and Levy, D.E. (2012) "A 2-phase labeling and choice architecture intervention to improve healthy food and beverage choices," *Am J Public Health*, **102**(3),527–33.

Townsend, M.S. (2010) "Where is the science? What will it take to show that nutrient profiling systems work?," *Am J Clin Nutr*, **91**(4), 1109S–1115S.

Van Camp, D.J. and Hooker, N.H. (2010) "Adoption of voluntary front of package nutrition schemes in UK food innovations," *Br Food J*, **112**(6), 580–591.

Van Herpen, E. and Van Trijp, H.C.M. (2011) "Front-of-pack nutrition labels. Their effect on attention and choices when consumers have varying goals and time constraints," *Appetite*, **57**, 148–160.

Van Herpen, E., Seiss, E. and Van Trijp, C.M. (2012) "The role of familiarity in front-of-pack label evaluation and use: a comparison between the United Kingdom and The Netherlands," *Food Qual Prefer*, **26**, 22–34.

van Kleef, E., Van Trijp, H., Paeps, F. and Fernández-Celemín, L. (2008) "Consumer preferences for front-of-pack calories labelling," *Public Health Nutr*, **11**(2), 203–213.

van Kleef, E. and Dagevos, H. (2013) "The growing role of front-of-pack nutrition profile labelling: a consumer perspective on key issues and controversies," *Crit Rev Food Sci Nutr*, DOI: 10.1080/10408398.2011.653018 (Accessed: 1 August 2014) (in press).

Verhagen, H. and van den Berg, H. (2008) "A simple visual model to compare existing nutrient profiling schemes," *Food Nutr Res*, **52**, doi: 10.3402/fnr.v52i0.1649.

Vyth, E.L., Steenhuis, I.H.M., Roodenburg, A.J.C., Brug, J. and Seidell, J.C. (2010) "Front-of-pack nutrition label stimulates healthier product development: a quantitative analysis," *Int J Behav Nutr Phys Act*, **7**, 65.

Vyth, E.L., Steenhuis, I.H.M., Heymans, M.W., Roodenburg, A.J.C., Brug, J. and Seidell, J.C. (2011) "Influence of placement of a nutrition logo on cafeteria menu items on lunchtime food choices at dutch work sites," *J Am Diet Assoc*, **111**, 131–136.

Vyth, E.L., Steenhuis, I.H.M., Brandt, H.E., Roodenburg, A.J.C., Brug, J. and Seidell, J.C. (2012) "Methodological quality of front-of-pack labeling studies: a review plus identification of research challenges," *Nutr Rev*, **70**(12), 709–720.

Watson, W.L., Kelly, B., Hector, D., Hughes, C., King, L., Crawford, J., Sergeant, J. and Chapman, K. (2014) "Can-front-of pack labelling schemes guide healthier food choices? Australian shopper's responses to seven labelling formats," *Appetite*, **72**, 90–97.

World Health Organization (WHO) (2010). *Guiding Principles and Framework Manual for the Development or Adaptation of Nutrient Profile Models* (1st edn.). Geneval: WHO. http://www.nutrinfo.com/biblioteca/libros_digitales/nuyrient.pdf, (Accessed: 1 August 2014).

Williams, P., McMahon, A. and Boustead, R. (2003) "A case study of sodium reduction in breakfast cereals and the impact of the *Pick the Tick* food information program in Australia," *Health Promot Int*, **18**(1), 51–55.

Williams, P., Duncan, R., de Agnoli, K., Hull, A., Owers, A. and Wang, T. (2010) "Front of pack daily intake labelling on Australian packaged foods: introduction and use 2007–2009," *Food Aust*, **62**, 583–588.

Yach, D., Lucio, A. and Barroso, C. (2007) "Can food and beverage companies help improve population health? Some insights from PepsiCo," *Med J Aust*, **187**(11–12), 656–657.

Young, L. and Swinburn, B. (2002) "Impact of the *Pick the Tick* food information programme on the salt content of food in New Zealand," *Health Promot Int*, **17**(1), 13–19.

Consumer interpretation of nutrition and other information on food and beverage labels

8

L. Lähteenmäki
MAPP University of Aarhus, Aarhus, Denmark

8.1 Introduction

The increase in obesity rates and related incidence of non-communicable diseases (WHO, 2011) has advanced the need to improve our diet and food-related behaviour. Since the quality of our diet is derived from products and meals that we purchase, there is public pressure to enable consumers to make informed, healthy choices when selecting their food. In the European Union (EU) this has led to legislation that requires provision of nutrition information in the food packaging (Regulation EU No 1169/2011). Providing nutrition information is mandatory but, in addition, food packaging can contain other information which is health-related, or is associated with health in consumers' minds. Mandatory nutrition information is typically placed on the back-of-package, whereas other health-related information is presented in the front-of-package (FOP) and is thereby more easily accessible to the consumers when purchasing.

The FOP information aims at simplifying nutrition- and health-related information by using standardised ways of expressing macronutrient and energy content of the product (e.g. guideline daily amounts (GDA), traffic lights), symbols that indicate the overall quality of the product in comparison to other products in the same product category, or claims with respect to specific nutrients or health outcomes. In the EU the legislation lists which nutrition and health claims are allowed (see http://ec.europa.eu/nuhclaims/), whereas the simplified nutrition labelling systems and symbols are based on voluntary labelling schemes which can be endorsed by authorities or by non-governmental organisations.

The health-related information differs in the level to which it provides direct guidance for the consumers (Hodgkins *et al.*, 2012). With plain nutrition information or nutrition claims, the food manufacturer offers consumers basic facts about the amount of nutrients, or whether the nutrient content is high or low in relation to other products. This leaves consumers with the task of processing that information and linking it to their existing knowledge in order to be able to judge whether the product would be good for them or not. This assessment requires combining information on several nutrients, or assessing nutrition claims in relation to other components in a

Advances in Food and Beverage Labelling. http://dx.doi.org/10.1533/9781782420934.2.133

product. Nutrient tables, or even graphic-assisted labelling, do not directly say what the health-related benefit of the product is. The same applies to nutrition claims; e.g. high calcium content can only be linked to beneficial bone health if consumers are aware of this link.

Health claims state what is the potential outcome of using the product. EC has special legislation covering the use of health claims (Regulation EC No 1924/2006) and a positive list of generic health claims that are allowed; most of these claims are related to nutrients and their functions in the body, and are based on existing, widely shared nutritional knowledge as presented in nutrition text books. There is also the possibility to apply for acceptance of claims based on new, emerging knowledge, but relatively few of these have been assessed by European Food Safety Authority (EFSA) as having the necessary level of scientific evidence backing the claims. There are a limited number of accepted risk-reduction claims which state a link between product use and reduction of a disease risk factor. Of claims approved to September 2013, the majority relate to lowering cholesterol levels and thus the risk of heart disease, or non-sugar sweetened chewing gums and a reduced risk of caries. However, for consumers, health claims should be easier messages to use in daily food purchases than nutrition information, as less effort is needed to process the information. Nutrition information is likely to be relevant to most people, whereas with health claims the target group profiting from the promised benefit can be more closely defined.

Health symbols that indicate the overall nutrition quality of the product provide direct guidance to the consumers. Health symbols are often based on nutrition profiles such as saturated fat, sugar, salt and/or fibre content (Schor *et al.*, 2010; Trichterborn*et al.*, 2011) so that products' overall quality, in comparison to the alternatives within the same product category, can be assessed. For consumers these symbols can act as direct recommendations for making better choices. The health symbols include the global "Healthy choices" which is a food industry based initiative (http://choicesprogramme.org/organisation), Swedish keyhole symbol (http://www.slv.se/en-gb/Group1/Food-and-Nutrition/Keyhole-symbol/), which has also been adopted in Denmark and Norway and is endorsed by food authorities. Other countries have similar programmes that can be driven by retailers, manufacturers or independent NGOs. Once consumers know what the symbol stands for, they can use that symbol as a direct guideline.

The nutrition-based information and symbols are not the only ones consumers use as cues when assessing the healthfulness of the product. Typically, organic production and labels are linked to increased perceived healthfulness (Magnusson *et al.*, 2003), although there is no consistent scientific evidence to back up this belief. Packages can also contain images and words that imply well-being and health. Active, running figures may imply that using the product will provide vigour to perform physical activity, and pictures of growing plants can be related to naturalness and increase the perceived healthiness of the product (Chrysochou and Grunert, 2013). Associations with naturalness have raised perceived healthfulness as natural is seen as innately better than anything that has been modified by man (Rozin *et al.*, 2004). Since processing implies modification, processed foods are seen as less healthy than non-processed foods (Evans *et al.*, 2010), and adding ingredients is seen as more unnatural than taking something out, even if the latter procedures can demand relatively advanced technologies (e.g. removing lactose from milk). Similarly to health claims and health symbols, consumers need to assess the relevance of organic or naturalness information to themselves. Sometimes the processing of this kind of information is unconscious, not requiring much effort.

This section has given an overview of the variety of health-related information that is available for consumers in the front-of-package (FOP) when making their choices. The next section describes how these different types of information are perceived and used by consumers, by looking separately at simplified nutrition information (GDAs, traffic lights, verbal information), health symbols, and health claims. That section is followed by an exploration of factors that influence consumers' ability and willingness to process health-related information in foods. The final section discusses future trends, knowledge gaps in our understanding of consumers' use of health-related information, and the needs for future research.

8.2 Consumer perception and use of front-of-package information

Nutrition- and health-related FOP information can be categorised according to its directiveness (Hodgkins *et al.*, 2012). The different types of information below are separated into three categories. Simplified nutrition information in FOP and nutrition claims both present non-directive or semi-directive information. Health symbols are openly directive, targeting the nutritional quality of the diet. Health claims are directive but promise specific health benefits that can be achieved from using single products.

8.2.1 Nutrition information and nutrition claims

Nutrition labelling on the product is basically providing information for consumers, but does not give consumers any directions towards better choices. Simplified FOP nutrition information contains nutrition claims which relate to high content of desirable nutrients (fibre, vitamins or minerals) or low content of undesirable nutrients (saturated fat, sodium or salt, saturated fat, energy), but the consumer needs to understand the relevance of the claimed nutrient in promoting health. Another form of nutrition labelling takes into account macronutrient contents, and gives a simplified profile of a product, in the form of GDA or as multiple traffic lights; the GDA basically gives information about nutrient content in relation to presumed amount an average person is supposed to ingest in a day, whereas the traffic light gives guidance as to whether the nutrient contents of macronutrients are good, reasonable or bad with green, amber and red colour coding, respectively.

There have been extensive studies on which format of nutrition information would be most useful to consumers when assessing the nutritional quality of a product. Some studies suggest that the traffic light system is more effective than

GDA in helping consumers to assess which product is more healthy (Borgmeier and Westenhoefer, 2009; Kelly *et al.*, 2009), but others have found no difference between the labelling systems, or that the results are country-dependent (Grunert *et al.*, 2010). The familiarity with the labelling systems has an impact on how well the different schemes are recognised, whether they are liked, and how easy it is to use them to make judgements between products (Grunert *et al.*, 2010; van Herpen *et al.*, 2012). The supporters of multiple traffic lights see the strong guidance element as a positive factor in the labelling system, especially for those consumers whose nutrition knowledge is low. GDA supporters assert that no products should be labelled as "bad" *per se*, as the question is about the amount eaten and frequency of use. Thus, the less guiding GDA gives tools to consumers to make their own decisions, without directly labelling the products as good or bad.

8.2.2 Health symbols

Health symbols indicate that the product is better than a conventional, average counterpart of the same product category. Use of health symbols is based on set rules which are endorsed by authorities, food industry or NGOs. For health symbols to be effective in guiding consumers' behaviour, consumers have to recognise the symbols, perceive and trust them in food products, and modify their behaviour accordingly.

Consumer awareness and recognition of the health symbols is rather high. The "Choices" logo was familiar to 62% of Dutch supermarket customers (Vyth *et al.*, 2010). The Scandinavian keyhole is recognised by 90% of Danish consumers, and 74% link it to a healthier choice, based on a report available in the Danish Food Safety Authorities webpages YouGov, 2012). Once consumers are aware of the logos, they are enabled to assess the healthiness of the products (Roberto *et al.*, 2012; Vyth *et al.*, 2009). Yet, there is very little evidence that the symbols would be directly used as guidance when making food choices (Roberto *et al.*, 2012; Steenhuis *et al.*, 2010).

One possible explanation has been that consumers do not notice the health symbols; thus, making them more visibly salient in food packaging would increase their use as well. One obvious way of increasing saliency is to increase the size of the symbols and position them centrally in the package (Orquin and Mueller Loose, 2013). When nutrition labels were presented in the cluttered part of the packaging they were not perceived as often as in a less-cluttered package area (Bialkova *et al.*, 2013; Visschers *et al.*, 2010). Similarly the amount of information as several pieces of health-related information reduced the visibility of nutrient labels (Bialkova et al., 2013). In their review on attention and decision making, Orquin and Mueller Loose (2013) point out that the most constant factor influencing consumers' attention when looking at packaging is based on goal orientation; hence, the health symbol needs to respond to consumers' interests. Any health-related symbol is likely to get more attention if consumers find health and good nutrition to be something they are interested in and which gives them useful information.

Roodenburg *et al.* (2011) showed with a simulation model that the average quality of diets would improve if consumers switched their choices to those that carry health symbols. The studies show that health symbols empower consumers' ability to make healthier choices, but yet the evidence supporting that use of health symbols as guiding factors in making choices is limited (Roberto *et al.*, 2012). However, health symbols may have a positive impact on our choices through the supply mechanism. Having more public emphasis on health in food choices may guide food manufacturers to develop new products that are able to carry the health symbols (Vyth *et al.*, 2010) and thus provide a wider selection and proportion of recommended options to the supermarket shelves. In an experimental set-up, participants made more healthy choices when the choice set was raised from 10 to 20 and included more healthful options to choose from. (Aschemann-Witzel *et al.*, 2013).

8.2.3 Health claims

Health claims give directive or semi-directive guidance to the consumers as they express health-related functions or health outcomes. In addition to scientific substantiation, the EU legislation requires that health claims need to be presented in a way that is understandable to an average consumer (Regulation EC No 1924/2006). Although health claims mainly considered are verbal statements designed for the FOP, the legislation also covers symbols and all marketing material related to products that carry health claims. As with other health-related information, consumers need to perceive the message and find it important enough for them to use the claim in their decision making. Due to the requirement of consumer understanding, health claims have been studied widely over the past 10 years, and there are recent reviews on consumer perceptions of health claims (Dean *et al.*, 2011; Lähteenmäki, 2013).

The health claims have mainly increased perceptions of the healthfulness of the product (Saba *et al.*, 2010; Sabbe *et al.*, 2009; van Trijp and van der Lans, 2007). However, this increase is often relatively small, and moderate at best. Thus, consumers also base their healthfulness evaluation on other product characteristics, and health claims do not override these other considerations. The type of product can play an important role: for example, bread and pasta are seen as more healthy than cakes (Saba *et al.*, 2010), and adding health claims to products that have a relatively low health image among consumers does not increase the perceived healthfulness of these products. Products with a low health image are often seen as unfit carriers for health claims (Krutulyte *et al.*, 2011). Sometimes the product with health claims is even perceived as less healthy (Lähteenmäki *et al.*, 2010); this may be due to inferences related to product naturalness, or the claimed effect is unfamiliar to the consumers so that they have no previous knowledge into which to link the claim.

Whether the wording of health claims has an impact on how consumers perceive health claims has been a central question. Health claims can contain a varying degree of information, depending whether they just mention the health outcome, or also refer to the active component and function on which the outcome is based (Grunert *et al.*, 2009). In some countries, the simple and short outcome claims are most appealing to consumers (Lynam *et al.*, 2011), or the results depend on the type of claim and benefit (van Kleef *et al.*, 2005; Wim Verbeke *et al.*, 2009). Cross-cultural studies indicate that perceptions of health claims vary widely between countries (Saba *et al.*, 2010; van Trijp and van der Lans, 2007). These results indicate that wording *per se* is of relatively little importance, and that the appeal and credibility of claims is highly context-dependent. Furthermore, consumers are not sensitive to qualifiers in the claims and make no distinction between claims that "provide" and "may provide" a benefit (Grunert *et al.*, 2009; Hooker and Teratanavat, 2008). Thus, claims that carry qualifiers expressing that the scientific evidence behind the claim is inconclusive or weak may be misleading for consumers. In the EU these kinds of claims are not allowed, whereas in the USA and Japan they are part of the market (Lalor and Wall, 2011).

Health claims can be framed either positively or negatively; a positive claim promises a benefit that can be achieved from the product use (e.g. stronger bones), whereas a negative claim would promise the avoidance of some negative consequence (e.g. helps to avoid the decrease in bone density). Typically risk-reduction claims are negative claims that promise a lower probability of declining health. Similarly to other wording related studies, whether a claim is positively or negatively framed seems to have no impact at all on consumer appeal (Grunert *et al.*, 2009), or the appeal is context-dependent (van Kleef *et al.*, 2005). Again, cultural factors may play a strong role here: for UK respondents, a short benefit claim was most attractive in cereal products, Finnish consumers liked a risk-reduction claim as well, and Italians preferred products with no claims at all (Saba *et al.*, 2010).

8.3 Consumers ability and motivation to process health-related information from food packages

The basic assumption behind providing consumers simplified FOP nutrition and health information is that it facilitates consumers' decision making, and hopefully results in better informed health choices. Informed, deliberate choices would require consumers to process the information and then decide whether the product is good for them or not. To make better choices, consumers have to have the opportunity, ability and motivation to make behavioural changes (Michie *et al.*, 2011). However, food choices are typically based on routine and low involvement, fast decision making. On average, consumers spend only a few seconds when purchasing products in the grocery store. This means that FOP health-related information can create an opportunity for consumers to make better choices, but this requires that the information is perceived, understandable and used for the aimed purpose. To achieve this, nutrition and health information have to compete for consumers' attention in order to be noticed. Although legislation and self-regulatory schemes can create the opportunity, the capability and motivation to use the information depend on factors that are person-, culture- and environment-dependent.

8.3.1 Consumer motivation

8.3.1.1 Socio-demographic factors

Women and older consumers tend to be more health-orientated in their food choices compared to men and younger consumers. Women have higher scores on the General Health Interest (GHI) scale (Roininen *et al.*, 1999, 2001) which measures respondents' interest in nutritionally healthy choices. The scale has been useful in predicting consumer interest in low-fat products (Urala and Lähteenmäki, 2007) and choices between healthy (apple) and unhealthy (chocolate bar) products (Roininen *et al.*, 2001), but not necessarily their willingness to choose products with health claims. The gender differences in interest towards products with health claims is less obvious than in nutritionally healthy products, and seem to depend on the type of benefit. Responses reflect actual risk levels of linked diseases: men are more positive about cholesterol-lowering products than women, whereas women rate benefits for bone health higher than men (Ares *et al.*, 2009; Dean *et al.*, 2007; Urala *et al.*, 2003). The responses to different benefits seem to vary between genders, but age seems to increase interest in all health-related information regardless of whether it is related to nutrition or specific benefits from health claims (Ares and Gámbaro, 2007; Cranfield *et al.*, 2011; Dean *et al.*, 2007; Herath *et al.*, 2008; Siegrist *et al.*, 2008).

Many of the responses to nutrition information are country-dependent. The way we consider healthiness in food and eating varies between cultures (Leeman *et al.*, 2011), and therefore the consumers in different countries differ in their responses. The importance and public awareness of various health-related issues differ between countries, which is reflected in the consumers' familiarity of the health-related information. When presented with GDA and multiple traffic light nutrition profiling in FOP, the respondents in UK and the Netherlands preferred the format of information they have been used to, although they could interpret the nutritional quality of the products with both schemes (van Herpen *et al.*, 2012). In Italy, any claims or logos with health-related connotations were seen negatively (Saba *et al.*, 2010). The cultural differences can be relatively strong, even in countries that have very similar product assortment and food culture: in a survey among the Nordic countries, Danish consumers were rather negative towards health claims, whereas Finnish and Swedish consumers saw them more positively (Grunert *et al.*, 2009). The likely reason for these differences between countries is previous exposure to similar kinds of health messages; in Denmark, health claims were not allowed before common EU legislation, whereas in Sweden and Finland consumers are used to these kinds of statements in their food products over several years of exposure in the market. Not only what is considered healthy is culturally dependent, but also what is considered unhealthy can have different meanings in different countries. When comparing responses to wholegrain products, countries did not differ from each other, but Italians rated refined grain products almost as healthy as wholegrain products, whereas Finnish respondents saw refined grain products as clearly unhealthy, and UK respondents were somewhere between these two countries in their assessments (Arvola *et al.*, 2007).

8.3.1.2 Health as a choice motivation in food products

When consumers list the reasons for their food choices, health is among the most frequently mentioned motivations (Margetts *et al.*, 1997; Renner *et al.*, 2012). Consumers vary in their attitudes towards health in foods, whether it is measured in relation to overall importance of health in food choices, nutritional healthiness which can be measure by the GHI scale (Roininen *et al.*, 1999), or attitudes towards products with health claims (Urala and Lähteenmäki, 2007). The GHI seems to be able to distinguish those more interested in low-fat products, whereas the attitudes towards functional foods can better explain the interest in cholesterol-lowering products. The strength of health awareness can modify responses to health-related information (Naylor *et al.*, 2009). Individuals with high health awareness seem to be less sensitive to conflicting health-related information than those whose health awareness is weak. Those with strong health awareness were less responsive to changes in information in an experimental study compared to those whose health awareness was weak.

Yet, health is not the only motivation, and not even necessarily the most important one, in our food choices. Pleasure derived from food (often expressed as good taste in consumer language), convenience, price and environmental concerns can play a role in our decision making. Adding health-related information to the FOP may emphasise the healthfulness of the product and thereby cause changes in the perception of other product attributes. Claiming that a product is healthy has been linked with a decrease in the expected tastiness (Raghunathan *et al.*, 2006), yet the association may be culture dependent, as in the French study healthiness was linked to tastiness and unhealthy food to bad taste (Werle *et al.*, 2013). Often health-related information is associated with modification of the product: adding components to achieve the health effect, or reducing or removing components to omit harmful effects. These modifications mean that the product has a certain degree of novelty, which may be perceived risky as such (Pliner and Hobden, 1992), or these modifications reduce the perceived naturalness (Evans *et al.*, 2010). This may be especially true for so-called functional foods that are marketed with health claims. In a study carried out in the Nordic countries (Iceland, Denmark, Finland, Norway and Sweden), adding omega-3 or bioactive peptides to a product was seen as a decrease in perceived naturalness for consumers, and the products became less attractive for them. The effect was also product-dependent, with the biggest effect on "unprocessed" pork meat compared to bread or yoghurt (Lähteenmäki *et al.*, 2010).

8.3.1.3 Perceived risk as a motivation to use health-related information

As described above, basic interest in nutritionally healthy eating is linked with consumers' motivation to process health-related information. However, the nutrition- and health-related information can be especially important when consumers perceive themselves at risk of a disease (Dean *et al.*, 2012) or they have a relative or family member at risk (Verbeke, 2005). The personal risk, though, seemed a much stronger factor in making products with risk-relevant health outcomes more appealing. In the study by Dean *et al.* (2012), consumers who rated themselves as having an increased risk of developing type 2 diabetes rated higher willingness to use a product that promised a risk reduction compared to those who did not consider themselves as being at risk, or having a relative or a close acquaintance who has increased risk.

8.3.2 Consumer capability to use health-related information

Consumers' capability to use nutrition- and health-related information in their food choices depends on their earlier knowledge and psychological capability of using the information when making choices. The psychological capability is dependent on other factors that go beyond having the right information (Bandura, 1998). Results from FOP nutrition labelling indicate that consumers are able to make the right conclusions about the products regardless of the form of presentation (Grunert *et al.*, 2010), although the multiple traffic light system seems to work better in some cases (Borgmeier and Westenhoefer, 2009; Kelly *et al.*, 2009). These results suggest that most consumers have the necessary ability.

Research on the relationship between nutrition knowledge and perception of health-related information indicates that understanding nutrition information from product labels is linked to better overall nutrition knowledge (Grunert *et al.*, 2010) in UK, but whether this also applies to the assessment of personal relevance is not known. A European study covering six countries shows that different demographic variables are linked to nutrition knowledge and interest in healthy eating (Grunert *et al.*, 2012). Nutrition knowledge does not differ between genders, but knowledge decreases with age, and health interest is clearly higher among women and increases with age (Grunert *et al.*, 2012). Nutrition knowledge is likely to improve consumers' capability of processing nutrition-related product information and thereby promote more elaborate and systematic processing. There is, however, a clear gap in the research as to whether basic nutrition knowledge is linked with understanding of health claims, both in the sense of understanding the benefit as promised by the claim and understanding its personal relevance.

8.4 Consumer understanding of health-related information

When interpreting health-related information on food packages, consumers have to rely on the information available and their existing knowledge. Interpreting health-related information requires a varying amount of effort depending on consumers' motivation, earlier experience, and capability. The so-called dual processing theories suggest that people have basically two routes to processing information: the elaborate, systematic way, which requires more working memory, resources, and effort; and the peripheral, heuristic processing, which is quick and based on cues that are easily available (Eagly and Chaiken, 1993; Petty and Cacioppo, 1986). The more effortful elaborate processing is likely to happen when individuals are highly motivated, find the information relevant for them, and are capable of processing the information. Typically, in studies where consumers are asked to assess products based on label information, the results are based on more effortful, elaborate, and systematic processing, whereas in everyday life food choices tend to be low involvement, fast, habitual decisions, which makes it likely that heuristic processing is dominant in most food choices. The systematic and heuristic processing can happen simultaneously, and heuristic cues can be elaborated in systematic processing.

Familiarity is closely linked to consumers' responses to health- and nutrition-related information. Familiar ways of presenting nutrition information based on known components make the use of information more appealing and more credible, as there are already existing reference points and knowledge on how to link the information. Familiarity in health claims makes them more convincing, as consumers can build on their existing belief structure (Lawson, 2002). In the Nordic study, the plain omega-3 fatty acid claims were regarded more positively when no information about their health benefits were linked to the claims, compared to the situation where benefits were mentioned as part of the claim (Lähteenmäki *et al.*, 2010). This odd result can be explained by the existing associative networks people have: mentioning plain omega-3 fatty acids may activate different nodes in these associative networks compared to linking the component to a specific benefit. Consumers are not just receivers of information when reading the nutrition-related information; they actively construct meanings to the messages by linking them to their existing knowledge. These linkages may create inferences that are objectively justified, based on the intended message, or unjustified, interpreting the messages beyond their content. With an unfamiliar nutrition or health message, these associative links do not exist and any beneficial piece of information is used to create new knowledge. However, with unfamiliar components, consumers tend to be cautious about novelty *per se* to a varying extent (Raudenbusch and Frank, 1999).

The health claim legislation requires that consumers are able to understand the claims – or the benefit of the claims. Grunert *et al.* (2011) introduced a method to measure understanding, whereby qualitative and quantitative approaches could be combined in a survey approach using realistic pictures of product packaging and other marketing material. The respondents were asked to describe, in their own words, what the benefits are and how they think the product works. The open-ended responses are categorised according to their content as safe, risky or vague. Safe answers describe the benefit clearly within what the claim promises, risky answers make assumptions beyond the claim content, and vague answers cannot be categorised as either safe or risky. In the study, 67% of respondents were categorised as giving safe responses, and 21% as giving risky responses. Although the share of consumers giving risky answers is only moderate, they were also the people with the most positive attitudes towards products with health claims, and thus the most likely to purchase these products.

Another aspect is the understanding of the relevance for oneself. Understanding the benefit as such does not tell whether the promised benefit would be useful to an individual. Products that promise to lower blood cholesterol levels are only beneficial when cholesterol levels are elevated, or if the consumer has a tendency to develop high blood cholesterol levels. The major part of accepted nutrition claims relate to normal development and maintenance of normal body functions that have relatively limited benefit value if the diet already contains sufficient amounts of these nutrients. Consumers may, however, mistrust the quality of their diet; whether this mistrust is justified or not, it may create uncertainty in consumers' minds. The research indicates that creating anxiety increases consumers' information search and use during choices (Hansen *et al.*, 2011), but this technique may not be the best way promote well-being.

8.5 Future trends

In order to make informed choices, consumers should have access to simplified, easily interpretable nutrition and health information at the point of purchase. This requires offering information on the front of packages, or alternatively information could be provided on shelves or via modern information technology. This also suggests that in the future European consumers will be bombarded with an increasing amount of nutrition and other health-related information on the food packages. In addition to nutrition labelling, the positive list of heath claims has been accepted by the EC and will be likely to bring more products carrying health claims to the market. For consumers this can mean an information overflow, wherein any piece of health-related information is drowned, not only among all the cues that are presented to convey other product attributes, but also among other health-relevant information. The increasing amount of health-related information can create a situation where health information becomes a norm that does not capture consumers' attention; instead of being additional beneficial information about the product quality, the health information can make it harder for the consumers to differentiate between products. If health information becomes a norm, it may have less direct influence on consumer behaviour than the behaviour of food suppliers who need to modify their new products so that they have positive nutrition profiles to present or fit into criteria set by symbols. The health claim legislation in the EU (Commission Regulation EU No 1160/2011) has a clause demanding that only products that fulfil a set nutrition profile are allowed to carry health claims. This clause should further protect consumers against claims that on the one hand promise a health benefit, but on the other hand have poor nutritional quality. Yet, as at the end of 2013, no decisions have been reached on defining the profiles.

Consumers seem to be able to use nutrition information in order to rank products according to their nutrition quality, and not interpret health claims as magic bullets providing benefits beyond what is promised in the claim. Whether consumers fully understand the nutritional quality, in the sense of how products contribute to the whole diet, remains unclear. With the increasing amount of health-related information about food supply, there should also be similarly increasing consumer education on basic nutrition knowledge and developing skills to support healthy eating. This would give better tools for consumers to process the product-specific information readily available. The nutrition education could be implemented as part of school curricula and include reading and analysing package labelling. Improved basic understanding of healthy eating could make consumers less vulnerable when assessing the personal relevance of different nutrition and health claims, but the education should also target improving consumers' skills to compile healthy diets.

Consumers' responses to nutrition labelling, health symbols, and nutrition and health claims have been studied largely in isolation from the rest of the package information. This gives us understanding as to how consumers perceive and understand the nutrition- and health-related information, but we do not know how other package design features interact and interfere with consumers' information processing. In a study situation, consumers are likely to respond to the given task of assessing products by using more elaborate and systematic processing of information than they use when making their routine daily food shopping. Therefore, more studies that simulate realistic purchase situations and product packaging are needed to see if the relatively optimistic view on consumers' ability to interpret the given information holds in the everyday shopping situation.

Nutrition- and health-related information provide an opportunity to make healthy choices. Consumers seem to be able to use that information when required to do so, but the crucial factor for making the effort is their motivation. If consumers are not interested in health when making their choices, labelling is likely to have no, or at best a very small, impact on their choice behaviour. Finding ways to reach the unmotivated consumers is a future challenge: how to improve their motivation to pay more attention to their eating habits, or how to change our environment so way that nutritionally sound options become the default choices. However, we should keep in mind that sometimes people who pay very little attention to the quality of their food choices may still have nutritionally good diets. On the other hand, highly motivated individuals may be those with atypical eating patterns and eating-related disorders. In addition to information, consumers need to acquire skills that enable producing healthful meals and other eating occasions from food products that are considered nutritionally sound. We need more research to improve our understanding of how consumers' interpretation of product quality is translated to diet quality.

References

Ares, G. and Gámbaro, A. (2007). Influence of gender, age and motives underlying food choice on perceived healthiness and willingness to try functional foods. *Appetite*, **49**(1), 148–158.

Ares, G., Giménez, A. and Gámbaro, A. (2009). Consumer perceived healthiness and willingness to try functional milk desserts. Influence of ingredient, ingredient name and health claim. *Food Quality and Preference,* **20**(1), 50–56.

Arvola, A., Lähteenmäki, L., Dean, M., Vassallo, M., Winkelmann, M., Claupein, E., Saba, A. and Shepherd, R. (2007). Consumers' beliefs about whole and refined grain products in the UK, Italy and Finland. *Journal of Cereal Science,* **46**(3), 197–206.

Aschemann-Witzel, J., Grunert, K. G., van Trijp, H., Bialkova, S., Raats, M. M., Hodgkins, C., Wasowicz-Kirylo, G.d, Koenigstorfer, J. *et al.* (2013). Effects of nutrition label format and product assortment on the healthfulness of food choice. *Appetite,* **71**, 63–74.

Bandura, A. (1998). Health promotion from the perspective of social cognitive theory. *Psychology and Health,* **13**, 623–649.
Bialkova, S., Grunert, K. G. and van Trijp, H. (2013). Standing out in the crowd: The effect of information clutter on consumer attention for front-of-pack nutrition labels. *Food Policy,* **41**, 65–74.
Borgmeier, I. and Westenhoefer, J. (2009). Impact of different food label formats on healthiness evaluation and food choice of consumers: A randomized-controlled study. *BMC Public Health,* **9**, 1–12.
Chrysochou, P. and Grunert, K. G. (2014). Health-related ad information and health motivation effects on product evaluations. *Journal of Business Research,* **67**(6), 1209–1217.
Cranfield, J., Henson, S. and Masakure, O. (2011). Factors affecting the extent to which consumers incorporate functional ingredients into their diets. *Journal of Agricultural Economics,* **62**(2), 375–392.
Dean, M., Lampila, P., Shepherd, R., Arvola, A., Saba, A., Vassallo, M., Claupein, E., Winkelmann, M. and Lähteenmäki, L. (2012). Perceived relevance and foods with health-related claims. *Food Quality and Preference,* **24**(1), 129–135.
Dean, M., Lähteenmäki, L. and Shepherd, R. (2011). Nutrition communication: Consumer perceptions and predicting intentions. *Proceedings of the Nutrition Society,* **70**(1), 19–25.
Dean, M., Shepherd, R., Arvola, A., Vassallo, M., Winkelmann, M., Claupein, E., Lähteenmäki, L., Raats, M. M. and Saba, A. (2007). Consumer perceptions of healthy cereal products and production methods. *Journal of Cereal Science,* **46**(3), 188–196.
Eagly, A. H. and Chaiken, S. (1993). *The Psychology of Attitudes.* Fort Worth, TX: Harcourt Brace College Publishers.
Evans, G., de Challemaison, B. and Cox, D. N. (2010). Consumers' ratings of the natural and unnatural qualities of foods. *Appetite,* **54**(3), 557–563.
Grunert, K. G., Fernandez-Celemin, L., Wills, J. M., Bonsmann, S. S. and Nureeva, L. (2010). Use and understanding of nutrition information on food labels in six European countries. *Journal of Public Health,* **18**(3), 261–277.
Grunert, K. G., Lähteenmäki, L., Boztug, Y., Martinsdottir, E., Ueland, Ø., Åström, A. and Lampila, P. (2009). Perception of health claims among Nordic consumers. *Journal of Consumer Policy,* **32**, 269–287.
Grunert, K. G., Wills, J., Celemín, L. F., Lähteenmäki, L., Scholderer, J. and Storcksdieck genannt Bonsmann, S. (2012). Socio-demographic and attitudinal determinants of nutrition knowledge of food shoppers in six European countries. *Food Quality and Preference,* **26**(2), 166–177.
Grunert, K. G., Wills, J. M. and Fernández-Celemín, L. (2010). Nutrition knowledge, and use and understanding of nutrition information on food labels among consumers in the UK. *Appetite,* **55**(2), 177–189.
Hansen, T., Mukherjee, A. and Thomsen, T. U. (2011). Anxiety and search during food choice: The moderating role of attitude towards nutritional claims. *Journal of Consumer Marketing,* **28**(3), 178–186.
Herath, D., Cranfield, J. and Henson, S. (2008). Who consumes functional foods and nutraceuticals in Canada? Results of cluster analysis of the 2006 survey of Canadians' Demand for Food Products Supporting Health and Wellness. *Appetite,* **51**(2), 256–265.
Hodgkins, C., Barnett, J., Wasowicz-Kirylo, G., Stysko-Kunkowska, M., Gulcan, Y., Kustepeli, Y., Akgungor, S., Chryssochoidis, G., Fernández-Celemin, L., Storcksdieck Genannt Bonsmann, S., Gibbs, M. and Raats, M. (2012). Understanding how consumers categorise nutritional labels: A consumer derived typology for front-of-pack nutrition labelling. *Appetite,* **59**(3), 806–817.

Hooker, N. H. and Teratanavat, R. (2008). Dissecting qualified health claims: Evidence from experimental studies. *Critical Reviews in Food Science and Nutrition,* **48**(2), 160–176.

Kelly, B., Hughes, C., Chapman, K., Louie, J. C. Y., Dixon, H., Crawford, J., King, L., Daube, M. and Slevin, T. (2009). Consumer testing of the acceptability and effectiveness of front-of-pack food labelling systems for the Australian grocery market. *Health Promotion International,* **24**(2), 120–129.

Krutulyte, R., Grunert, K. G., Scholderer, J., Lähteenmäki, L., Hagemann, K. S., Elgaard, P., Nielsen, B. and Graverholt, J. P. (2011). Perceived fit of different combinations of carriers and functional ingredients and its effect on purchase intention. *Food Quality and Preference,* **22**(1), 11–16.

Lalor, F. and Wall, P. G. (2011). Health claims regulations. Comparison between USA, Japan and European Union. *British Food Journal,* **113**(2), 298–313.

Lawson, R. (2002). Consumer knowledge structures: Background issues and introduction [Article]. *Psychology and Marketing,* **19**(6), 447–455.

Leeman, R. F., Fischler, C. and Rozin, P. (2011). Medical doctors' attitudes and beliefs about diet and health are more like those of their lay countrymen (France, Germany, Italy, UK and USA) than those of doctors in other countries. *Appetite,* **56**(3), 558–563.

Lynam, A. M., McKevitt, A. and Gibney, M. J. (2011). Irish consumers' use and perception of nutrition and health claims. *Public Health Nutrition,* **14**(12), 2213–2219.

Lähteenmäki, L. (2013). Claiming health in food products. *Food Quality and Preference,* **27**(2), 196–201.

Lähteenmäki, L., Lampila, P., Grunert, K., Boztug, Y., Ueland, Ø., Åström, A. and Martinsdóttir, E. (2010). Impact of health-related claims on the perception of other product attributes. *Food Policy,* **35**(3), 230–239.

Magnusson, M. K., Arvola, A., Hursti, U. K. K., Aberg, L. and Sjoden, P. O. (2003). Choice of organic foods is related to perceived consequences for human health and to environmentally friendly behaviour. *Appetite,* **40**(2), 109–117.

Margetts, B. M., Martinez, J. A., Saba, A., Holm, L. and Kearney, M. (1997). Definitions of 'healthy' eating: A pan-EU survey of consumer attitudes to food, nutrition and health. *European Journal of Clinical Nutrition,* **51**, 23–29.

Michie, S., van Stralen, M. M. and West, R. (2011). The behaviour change wheel: A new method for characterising and designing behaviour change interventions. Implementation Science, 6(1), Article 42.

Naylor, R. W., Droms, C. M. and Haws, K. L. (2009). Eating with a purpose: Consumer response to functional food health claims in conflicting versus complementary information environments. *Journal of Public Policy and Marketing,* **28**(2), 221–233.

Orquin, J. L. and Mueller Loose, S. (2013). Attention and choice: A review on eye movements in decision making. *Acta Psychologica,* **144**(1), 190–206.

Petty, R. E. and Cacioppo, J. T. (1986). The elaboration likelihood model of persuasion. *Advances in Experimental Social Psychology,* **19**, 123–205.

Pliner, P. and Hobden, K. (1992). Development of a scale to measure the trait of food neophobia in humans. *Appetite,* **19**, 105–120.

Raghunathan, R., Naylor, R. W. and Hoyer, W. D. (2006). The unhealthy = Tasty intuition and its effects on taste inferences, enjoyment, and choice of food products. *Journal of Marketing,* **70**(4), 170–184.

Raudenbusch, B. and Frank, R. A. (1999). Assessing food neophobia: The role of stimulus familiarity. *Appetite,* **32**(2), 261–271.

Regulation (EC) No 1924/2006 of the European Parliament and of the Council of 20 December 2006 on nutrition and health claims made on foods. *Official Journal of the European Union*, L 404, 9–25.

Regulation (EU) No 1169/2011 of the European Parliament and of the Council of 25 October 2011 on the provision of food information to consumers. *Official Journal of the European Union*, L304/18.

Renner, B., Sproesser, G., Strohbach, S. and Schupp, H. T. (2012). Why we eat what we eat. The Eating Motivation Survey (TEMS). *Appetite*, **59**(1), 117–128.

Roberto, C. A., Shivaram, M., Martinez, O., Boles, C., Harris, J. L. and Brownell, K. D. (2012). The Smart Choices front-of-package nutrition label. Influence on perceptions and intake of cereal. *Appetite*, **58**(2), 651–657.

Roininen, K., Lähteenmäki, L. and Tuorila, H. (1999). Quantification of consumer attitudes to health and hedonic characteristics of foods. *Appetite*, **33**, 71–88.

Roininen, K., Tuorila, H., Zandstra, E. H., de Graaf, C., Vehkalahti, K., Stubenitsky, K. and Mela, D. J. (2001). Differences in health and taste attitudes and reported behaviour among Finnish, Dutch, and British consumers: A cross-national validation of the health and taste attitude scales (HTAS). *Appetite*, **37**(1), 33–45.

Roodenburg, A. J. C., Schlatmann, A., Dötsch-Klerk, M., Daamen, R., Dong, J., Guarro, M., Stergiou, M., Sayed, N., Ronoh, E., Jansen, L. and Seidell, J. C. (2011). Potential effects of nutrient profiles on nutrient intakes in the Netherlands, Greece, Spain, USA, Israel, China and South-Africa. *PLoS ONE*, **6**(2), 2011 Article e14721.

Rozin, P., Spranca, M., Krieger, Z., Neuhaus, R., Surillo, D., Swerdlin, A. and Wood, K. (2004). Preference for natural: Instrumental and ideational/moral motivations, and the contrast between foods and medicines. *Appetite*, **43**(2), 147–154.

Saba, A., Vassallo, M., Shepherd, R., Lampila, P., Arvola, A., Dean, M., Winkelmann, M., Claupein, E. and Lähteenmäki, L. (2010). Country-wise differences in perception of health-related messages in cereal-based food products. *Food Quality and Preference*, **21**, 385–393.

Sabbe, S., Verbeke, W., Deliza, R., Matta, V. and Van Damme, P. (2009). Effect of a health claim and personal characteristics on consumer acceptance of fruit juices with different concentrations of açaí (Euterpe oleracea Mart.). *Appetite*, **53**(1), 84–92.

Schor, D., Maniscalco, S., Tuttle, M. M., Alligood, S. and Kapsak, W. R. (2010). Nutrition facts you can't miss: The evolution of front-of-pack labeling: Providing consumers with tools to help select foods and beverages to encourage more healthful diets. *Nutrition Today*, **45**(1), 22–32.

Siegrist, M., Stampfli, N. and Kastenholz, H. (2008). Consumers' willingness to buy functional foods. The influence of carrier, benefit and trust. *Appetite*, **51**(3), 526–529.

Steenhuis, I. H. M., Kroeze, W., Vyth, E. L., Valk, S., Verbauwen, R. and Seidell, J. C. (2010). The effects of using a nutrition logo on consumption and product evaluation of a sweet pastry. *Appetite*, **55**(3), 707–709.

Trichterborn, J., Harzer, G. and Kunz, C. (2011). Nutrient profiling and food label claims: Evaluation of dairy products in three major European countries. *European Journal of Clinical Nutrition*, **65**(9), 1032–1038.

Urala, N., Arvola, A. and Lähteenmäki, L. (2003). Strength of health-related claims and their perceived advantage. *International Journal of Food and Technology*, **38**, 815–826.

Urala, N. and Lähteenmäki, L. (2007). Consumers' changing attitudes towards functional foods. *Food Quality and Preference*, **18**(1), 1–12.

Van Herpen, E., Seiss, E. and Van Trijp, H. C. M. (2012). The role of familiarity in front-of-pack label evaluation and use: A comparison between the United Kingdom and The Netherlands. *Food Quality and Preference*, **26**(1), 22–34.

van Kleef, E., van Trijp, H. C. M. and Luning, P. (2005). Functional foods: Health claim-food product compatibility and the impact of health claim framing on consumer evaluation. *Appetite,* **44**(5), 299–308.

van Trijp, H. C. M. and van der Lans, I. A. (2007). Consumer perceptions of nutrition and health claims. *Appetite,* **48**(3), 305–324.

Verbeke, W. (2005). Consumer acceptance of functional foods: Socio-demographic, cognitive and attitudinal determinants. *Food Quality and Preference,* **16**(1), 45–57.

Verbeke, W., Scholderer, J. and Lahteenmaki, L. (2009). Consumer appeal of nutrition and health claims in three existing product concepts. *Appetite,* **52**(3), 684–692.

Visschers, V. H., Hess, R. and Siegrist, M. (2010). Health motivation and product design determine consumers visual attention to nutrition information on food products. *Public Health Nutrition,* **13**(7), 1099–1106.

Vyth, E. L., Steenhuis, I. H. M., Mallant, S. F., Mol, Z. L., Brug, J., Temminghoff, M., Feunekes, G. I., Jansen, L., Verhagen, H. and Seidell, J. C. (2009). A front-of-pack nutrition logo: A quantitative and qualitative process evaluation in the Netherlands. *Journal of Health Communication,* **14**(7), 631–645.

Vyth, E. L., Steenhuis, I. H. M., Roodenburg, A. J. C., Brug, J. and Seidell, J. C. (2010). Front-of-pack nutrition label stimulates healthier product development: A quantitative analysis. *International Journal of Behavioral Nutrition and Physical Activity,* **7**, 65. http://www.ijbnpa.org/content/7/1/65.

Werle, C. O. C., Trendel, O. and Ardito, G. (2013). Unhealthy food is not tastier for everybody: The "healthy = tasty" French intuition. *Food Quality and Preference,* **28**(1), 116–121.

WHO (2011). Global status report on noncommunicable diseases. Report downloaded from whqlibdoc.who.int/publications/2011/9789240686458_eng.pdf 9 January, 2012.

YouGov (2012). http://www.noeglehullet.dk/NR/rdonlyres/178C2541-115C-407B-8E70-9BA46522A1F9/0/YouGov_oktober_2012.pdf. Downloaded on 23.9.2013.

Part Three

Trends in labelling relating to other aspects of food quality

Ethical and environmental labelling of foods and beverages

9

J. Troth
Assessment Compliance Expertise Ltd, Lichfield, UK

9.1 Introduction

Over the last two to three centuries, the developed countries have moved from a rural-based economy to urban- and industrial-based economies, resulting in large population moves from the countryside, where their food was produced and traded locally, to towns and cities, which rely upon bringing in large quantities of food from surrounding farming areas. During this urbanisation and industrialisation, the distance between food production and food consumption has grown. Coupled with this, and with increasing wealth generated by industry and increases in tourism, consumers want a much more varied and exciting range of food stuffs, sourced from around the globe. The growing distance between production and consumption has detached the connection between consumers and producer, with ever-increasing intermediaries such as commodity traders, manufacturers, wholesalers, distributors and retailers.

The increasing demand to supply in this market has led to an industrialisation of agriculture and many food products becoming tradeable commodities, leading to more intensive production methods such as indoor farming systems, heavy use of agrochemicals and the exploitation of small scale producers in the developing world, by powerful multi-national companies in the developed world.

As food moved from being a basic necessity to being available in abundance, the cost of food has dropped relative to household incomes. People subsequently became aware of the impact that this was having on agriculture and producers in developing countries. In order to produce food plentifully and cheaply, the environment, animals and farmers were paying a higher price.

As consumer purchasing power in the developed world has grown, and there is plentiful supply of food, this has given many consumers the option to make food purchasing decisions based on "lifestyle choice" rather than "basic need of nutrition." The last 20–30 years have seen a proliferation of labelling schemes, with the aim of conveying to the consumers how the food was produced so that they can purchase according to their lifestyle choices.

This chapter aims to inform the reader of what defines ethical and environment labelling schemes, what they aim to deliver, and how they are controlled and

Advances in Food and Beverage Labelling. http://dx.doi.org/10.1533/9781782420934.3.151

monitored. It will discuss consumer expectations and their motivation for making such purchases and describe the main global schemes and a summary of industry and single issue initiatives.

9.2 Defining ethical labelling

Ethical labelling covers a wide range of criteria in food production and purchasing choice. Ethical purchases are generally based on the personal values of the consumer making the purchasing decision, and so ethical labelling initiatives have to clearly identify what values the product or logo supports. Values are generally based on a sense of "fairness" and "justice" and doing the right thing. Examples of values include ensuring humans are treated fairly (social justice), animals are treated humanely (animal welfare), protection of natural habitats and that the environment is respected.

The Food Ethics Council of Great Britain defines ethics as referring to the values, principles and codes by which people live (http://www.foodethicscouncil.org/topic/Ethics) (accessed 15 June 2013). Acting ethically means taking values seriously and asking "what should I do, all things considered?" Advice on making better decisions based on ethical principles include: working out the winners and losers, asking who benefits most from your decision, being clear and open about your principles and understanding the benefits of talking difficult decisions through. A report of a business forum published by the Food Ethics Council (2009) reports on a Business Forum Meeting on 20 January 2009 covering ethical consumption, consumer cultures and resource constraints.

However, these questions cover a wide range of issues and, in the purchasing environment, consumers are having to make hundreds of purchasing decisions based on what they require, product value, appeal, health needs and, of course, cost – so when ethical choices are included in the purchasing decision mix, it adds another tier of decisions. For example, local versus fairly traded, free range versus organic, dolphin friendly versus sustainably fished. This array of ethical choices often leads to pressure and confusion. In order to make ethical purchasing choices, consumers have to be well informed about the values that appeal to them and how certain products and schemes deliver those values. Often the choice is made at retailer or caterer level, for example by offering only ethical or sustainably produced products, or supplying "own label" ranges that are fairly traded or organically produced.

Ethical labelling therefore can cover a wide range of issues, some of which cross over into environmental labelling. For development of ethical labelling schemes, there are limited regulatory or international standards. Over the years, a range of schemes have developed which address ethical concerns. These can be product specific – for example, free range eggs – or cover a wide range of products internationally, such as Fairtrade. There are a wide range of established ethical labelling schemes covering animal welfare, social justice and environmental protection and some will be discussed in more detail in later chapters. Some are single issue labels such as "dolphin friendly," while others cover a wide range of ethical issues, such as sustainability, social and economic criteria.

9.3 Defining environmental labelling

For environmental labelling, it would be impossible to have a single definition, due to the massive range of factors that "the environment" covers, such as protection of the environment, greenhouse gas output, water sustainability and reduction in the use of non-renewable resources. Other considerations include whether an environmental benefit is defined at local level, such as farming methods, or on a much wider definition based on "life-cycle" analysis which includes the environmental impact of production, processing and final use. One example is the paradox of consumers trying to assess the environmental benefits of "environmentally friendly" products that are flown half way around the world, versus intensively farmed products that are produced locally, therefore with minimal transport impact. However, the International Standards Organisation (ISO) have developed standards for environmental labelling criteria and an international, non-profit organisation called the Global Ecolabelling Network (GEN) was established in 1994 to ensure mutual recognition and common core performance criteria for ecolabelling initiatives. The mission of GEN is to (http://www.globalecolabelling.net/about/what_is_gen/mission/index.htm), (accessed 27 May 2013):

- Serve its members, other ecolabelling programmes and stakeholders, and the public by improving, promoting and developing the ecolabelling of products and the credibility of ecolabelling programmes worldwide;
- Foster cooperation, information exchange and harmonisation among members, associates and other ecolabelling programmes;
- Participate in international organisations in order to promote ecolabelling generally; and
- Encourage the demand for, and supply of, more environmentally responsible goods and services.

In support of this mission, GEN members:

- Set criteria for and certify products and services with lower environmental burdens and impacts than comparable products/services;
- Provide information, advice and technical assistance to organisations contemplating or developing programmes;
- Disseminate information to the public; and
- Represent the interests of ecolabelling in various international meetings and events.

The ISO standards and the GEN are primarily aimed at non-food product ecolabels (e.g. electrical appliances), although some food products are covered by GEN programmes – these include wine (Australia), coffee (North America), soft drinks (China). The GEN published an information paper in 2004 entitled "Introduction to Ecolabelling" (GEN, 2004). The information paper defines an ecolabel as:

> *...a label which identifies overall environmental preference of a product (i.e. good or service) within a product category based on life-cycle considerations. In contrast to a self-styled environmental symbol or claim statement developed by a manufacturer or service provider, an ecolabel is awarded by an impartial third party to products that meet established environmental leadership criteria.*

As has been identified by the International Organization for Standardization (ISO), the overall goal of these labels and declarations is (GEN, 2004):

> *...through communication of verifiable and accurate information, that is not misleading, on environmental aspects of products and services, to encourage the demand for and supply of those products and services that cause less stress on the environment, thereby stimulating the potential for market-driven continuous environmental improvement.*

The ISO has attempted to standardise the principles, practices and key characteristics relating to three major voluntary environmental labelling types: Type I – environmental labelling (i.e. ecolabel), Type II – self-declaration claims and Type III – environmental declarations (e.g. report cards/information labels)

1. ISO 14024:1999: Ecolabelling schemes, environmental labels and declarations – Type I environmental labelling – Principles and procedures.
2. ISO 14021:1999: Self-declared environmental claims, Environmental labels and declarations – Type II environmental labelling.
3. ISO 14025:2006: Life-cycle data declarations: Environmental labels and declarations – Type III environmental declarations – Principles and procedures.

The ISO standards cover ecolabelling schemes in depth; however, a convenient summary guide has been published by ISO (2012) which gives background to the development of the guide and a description of the requirements of each of the standards. For example, the reasons for standards, requirements for claims, selection of environmental criteria, product categories, verification methods, mutual recognition criteria and the use of symbols.

There are few legal definitions or controls over ecolabelling claims, other than trades descriptions, consumer protection legislation and compliance of products with environmental regulations relating to emissions or pollution. Also, it is often very difficult to quantify the environmental benefits of certain methods of production. Using organic farming as an example, it is probable that not spraying pesticides or applying fertilisers should improve the environment, but if more cultivation has to be done to control weeds, then the pay-off is more fuel consumption and contribution to global warming. Numerous studies have been done on environmental impact of different farming methods, for example FAO (2011), Foster *et al.* (2006) and Venkat (2012).

9.4 Control of labelling schemes

Although there has been massive growth in the range and type of labelling claims, legislation has not kept pace and the various claims remain largely unregulated. Some schemes, such as organic food and farming, have become part of a regulated system in a number of countries. However, due to the often complex and cross-border supply chains, the ability to regulate schemes is limited. Brand reputation and the desire to maintain consumer confidence have been the biggest factors in ensuring scheme compliance. The new EU Food Information Regulation EU1169/2011 makes a couple of references to provision of information to protect consumers' health and interests and to enable consumers to make informed choices with regard to health, economic, environmental, social and ethical considerations (Chapter II, Article 3, Para 1) and an acknowledgement that consumers choices can be influenced by health, economic, environmental, social and ethical considerations; however, there is little other reference in the regulation to these criteria.

The various claims are largely regulated by trading and business law, such as trademark use and law of contract. Most standards and schemes are industry-or non-governmental organisation (NGO)-developed, although many of the schemes are now independently verified by certification bodies and their "impact" (effectiveness) monitored.

To help readers who may not be familiar with certification terminology, a brief overview is given below.

9.4.1 Private/self-certification schemes

These are schemes where the producers themselves, non-governmental bodies, trade associations or retail companies establish standards and undertake the assessment and monitoring of the standards and use of the logo.

9.4.2 Third party certification

These are independent certification bodies that have no relationship with the companies they audit and certify, or with the standards owners. They are usually contracted to supply certification and assessment services to standards-setting organisations, and usually operate to internationally recognised standards such as ISO via accreditation by national or international accreditation organisations.

9.4.3 Accreditation

Accreditation is a system of monitoring certification bodies to ensure they are competent, independent and consistent in their approach. Accreditation bodies usually assess against ISO, European Norm (EN) or similar international standards, and can be either national organisations such as United Kingdom Accreditation Service (UKAS) or international independent organisations such as the International Organic Accreditation Service (IOAS).

9.4.4 Regulation

Only the organic/biological system of production and product labelling is regulated in various countries. The European organic regulation and most other organic regulations require that certification bodies are accredited to EN45011/ISO65 by their national accreditation bodies and that certification bodies are approved by competent authorities (e.g. agricultural or food ministries) in the member state in which they operate. There have been a number of examples of prosecutions, some leading to imprisonment, where fraudulent activity in relation to organic food production and labelling has been identified.

There are a number of environmental and ethical organisations which have been formed to ensure that labelling schemes are based on sound standards. One leading organisation is the International Social and Environmental Accreditation and Labelling (ISEAL) Alliance initiated in 2002 by four founding organisations – International Federation of Organic Agricultural Movements (IFOAM), Fairtrade, Forest Stewardship Council and Marine Stewardship Council. The ISEAL Alliance is an association of leading voluntary international standard setting and conformity assessment organisations that focus on social and environmental issues. ISEAL Alliance members collaborate to build international recognition and legitimacy for their programmes. The Alliance includes members such as Fairtrade, Rainforest Alliance, Roundtable on Sustainable Palm Oil (RSPO), Marine Stewardship Council, UTZ, Bonsucro and the IOAS. The four goals of ISEAL Alliance are to (http://www.isealalliance.org/about-us) (accessed 15 June 2013):

- Demonstrate and improve impacts of standards
- Improve the effectiveness of standards
- Define credibility for sustainability standards
- Increase the adoption of credible sustainability standards

In recent years, various codes have been established, such as a standards-setting code and an impacts code – the latter provides a process for the effective measurement of standards systems and evidence of their contribution to social and environmental impacts on the ground, with the aim of further developing transparency, consumer trust and confidence that their purchasing choice, using involving a price premium, is making a difference. Further information on the work and role of ISEAL can be found on their website: www.isealalliance.org.

9.5 The motivation behind consumer purchasing choices

The use of "consumer power" to change an undesirable practice is not a recent idea. One of the first documented examples dates back to 1791 at the time when slavery was being used to supply the demands of Europe and America for sugar, tobacco and cotton. British campaigners who had become aware of the brutality and human suffering of this practice tried to get it abolished by the government, but the abolition bill was rejected by Parliament. The campaigners realised that the slave trade was economically dependent on the sale of the products of slavery, namely sugar, tea, coffee, rum and tobacco. An "anti-sugar" pamphlet was written by William Fox and published in 1791. It sold around 70,000 copies in 4 months and by 1792, around 400,000 consumers were boycotting slave grown sugar. Although the product itself probably was not labelled as "slave-free," it is an early example of informing consumers about production methods and of the consumers using their purchasing power to influence change. This example displays the same consumer characteristics that are applicable to today's environmental and ethical labelling schemes – informed consumers using their purchasing power to support or discourage particular practices.

There are a wide range of factors that influence consumers to make environmental and ethical purchases, and the purchasing choice is often based on a complex set of factors. These will include:

- how well informed the consumer is
- how interested they are in their purchasing choices
- what value they place on food – i.e. how much they are prepared to pay

The factors influencing purchasing choice can be broken down into two broad types – those that have a personal benefit, and those which are more altruistic.

Personal benefits include:

- health – avoidance of pesticides, veterinary residues
- improved taste, better quality
- safety
- price

Wider altruistic reasons which do not impart personal benefits include:

- animal welfare
- protection of habitats/environment
- social justice

The rapid pace of change in science and technology, changes in legislation and the current socio-economic and socio-demographic realities have all had a marked impact on the food we buy today. There have been numerous studies that assess factors that influence purchasing behaviour with regard to environmentally-and ethically-labelled foods. The studies reviewed indicate that personal benefits rank higher than altruistic benefits, when purchasing eco-products or ethical products. One such study was by the UK-based non-profit consumer organisation "Which" that published a report in 2010 entitled "Making sustainable food choices easier." Their research had shown that even informed consumers struggle to understand most of the environmental and ethical labelling schemes on the market. The study noted that three out of four consumers who were surveyed believed that protecting the environment is important when making their food purchasing choice. The report highlighted consumer views that should inform the development of such schemes based on the following principles:

- Simplicity: ensure labels communicate clear, short messages.
- Impact: make sure labels are easily noticeable and understandable, as few people have the motivation to seek them out.
- Consistency: avoid too many different schemes or different positioning on the label.
- Coherence: explore the extent to which issues can be combined into a composite label.
- Evidenced: ensure the criteria that underpin the scheme that reflect the available scientific evidence.
- Independence: ensure schemes are independent.

The report then goes on to show which factors are most important to consumers when purchasing food. The most important, at 76%, was taste and quality; second was safety, at 72%; third was price, at 55%; healthy eating was fourth at 50%;

animal welfare was fifth, protecting the environment sixth and organic was last, at 11%. This demonstrates the confusion over consumer understanding of organics, as organic agriculture helps to protect the environment and has the highest welfare standards of all production systems.

Likewise a survey, requested by the European Commission (2010) on food-related risks, identified such consumer concerns as the intensification of farming, the use of pesticides, the industrialisation of food production, using additives and preservatives to improve taste, appearance and shelf-life, for example. Furthermore, new technologies such as genetically modification (GMO), nanotechnology and animal cloning are altering agriculture and food production today. The survey was carried out on a representative sample of about 27,000 individuals, aged 15 or over in all 27 member states. Pesticide residues in fruit, vegetables and cereals are the overall highest concern in 12 member states, confirming its top position of concern with EU citizens. This issue is of particular concern in Greece (91%), Lithuania (88%), Italy (85%), Luxembourg (85%), Bulgaria (84%), Hungary, (84%), France (80%), Malta (77%), Slovenia (75%), Germany (75%), Belgium, (72%) and Austria (67%).

In other northern European countries, including Sweden (74%), Finland (66%) and the UK (67%), it is "the welfare of farmed animals" which arouses the highest level of concern among citizens. Concerns such as "additives like colours, preservatives or flavouring used in food or drinks" and "residues like antibiotics or hormones in meat" are each the largest sources of worry in countries such as Poland (79%), Romania (70%), Cyprus (92%) and the Netherland (63%). It must be noted that the survey was based on consumer concerns relating to a range of food hazards and risks, and not just on how food was labelled. However, it is those concerns relating to personal safety and risk that are highlighted. Concerns over the environment rated only 3% with those questioned.

A study cited in the Soil Association's Organic Market Report (2012) identified the following reasons why consumers did or did not purchase organic foods (Table 9.1).

Table 9.1 Motivation for buying or not buying organic food

Top reasons for buying organic food	%	Reasons for not buying organic food	%
Fewer chemicals	62	Organic food is too expensive	91
Natural and unprocessed	57	Organic food is no healthier	21
Healthier for me and my family	52	The range of products is not broad enough	21
Better for nature and the environment	47	I don't know enough about it	15
Organic food tastes better	44	I can't always tell if it is organic or not	13
Safer to eat	41	There is no nutritional benefit	13
Organic farming is kinder to animals	34	I'm not the sort of person who buys organic	13
More ethical	33		
No GM ingredients	29		
There are good offers	13		

Source: Soil Association, 2012

Again, this demonstrates that "personal" benefits (e.g. healthier/fewer chemicals) are placed ahead of wider environmental or ethical concerns, such as environment or welfare. This is only human nature, but it is an important point to bear in mind when labelling initiatives are being considered, that to encourage "consumer buy in," some personal benefits for the purchasing the product (e.g. better flavour/safety) need to considered.

9.6 A review of global environmental and ethical schemes

The following labelling schemes are internationally recognised and will be reviewed in terms of what they aim to deliver, how they are monitored and example size of market.

9.6.1 Organic food and farming

Organic farming (also referred to as biological "bio" or ecological "eco") is the most widespread of the environmental and ethical labelling initiatives in terms of area and value. Originally intended as a way of producing food sustainably, by using rotation and recycling of nutrients with minimal reliance on brought-in inputs, organic farming has developed over the years to include a broader range of criteria including animal management and welfare, and conveying the benefits of organic farming into processed consumer products.

The first description of "biological" or "biodynamic agriculture" was in a series of lectures by the Austrian philosopher Rudolph Steiner in 1924. In the UK in 1940, Sir Albert Howard published "An Agricultural Testament," advocating that Britain preserve the "cycle of life" and adopt "permanent agriculture" systems, using urban food waste and sewage to build soil fertility. In 1946, a group of likeminded people, including Howard and led by Eve Balfour, founded the Soil Association in the UK. The primary objective of the Soil Association was to encourage research and development in organic farming methods to counter increasing intensification of farming methods to overcome the shortages of food following WWII.

Interest in organic farming grew throughout Europe and the USA during the 1960s and, as a result, trade increased and the production, processing and trade became regulated to formalise and harmonise production, processing and labelling methods begin to ensure consumer confidence in the system and protection of the industry. This has led to organics being one of the most formalised and regulated environmental labelling systems. Control bodies in the EU have to be approved by government authorities, be accredited to EN45011/ISO17065 standards, and all operators are subject to annual inspection. The whole chain, from farming to processing, including point of sale if goods are sold loose, is regulated. Animal feed and seed packers, storage and wholesale facilities also require certification and inspecting.

In order to ensure consistency in standards to facilitate world trade, a group of organic organisations came together in 1972 to form the International Federation of Organic Agricultural Movements (IFOAM). This created a set of guiding principles on which organic standards around the world should be based. These are (http://infohub.ifoam.org/en/what-organic/principles-organic-agriculture) (accessed 13 June 2013):

- The principle of health
- The principle of ecology
- The principle of fairness
- The principle of care

Codex Alimentarius also formulated a set of organic standards which are used as base criteria for a number of worldwide standards. As organic agriculture developed, its scope and remit has widened to include animal welfare, social and ethical standards, broadening into non-food areas such as textiles and body care products.

Environmental criteria that the European Commission identify with organic agriculture are (http://ec.europa.eu/agriculture/organic/environment_en) (accessed 2 June 2013):

- Protection of soil
- Maintaining biodiversity
- Better water retention by improved soil structure
- Prevention of ground water contamination

The animal welfare criteria that the European Commission identify with organic agriculture are (http://ec.europa.eu/agriculture/organic/animal-welfare_en) (accessed 2 June 2013):

- Animal production should be outdoors-and pasture-based, when housed due to climatic conditions, then appropriate bedding and litter must be provided,
- Intensive systems and tethering are prohibited,
- Mutilations such as removal of tails and beaks are heavily restricted,
- Limits on flock sizes for poultry and stocking levels,
- Feed should be organically produced, natural and not contain animal by-products medications, growth promoters or genetically modified organisms (GMO) materials,
- Youngstock should be fed on natural milk, ideally from their mothers.

Processing criteria are (http://ec.europa.eu/agriculture/organic/consumer-confidence/food-quality_en) (accessed 2 June 2013):

- Restricted range of food additives including prohibition of colours and very limited use of preservatives (e.g. only nitrates permitted for curing of meat only),
- Prohibition of genetically modified ingredients,
- Very restricted list of permitted non-organic ingredients that are not yet available as organic.

9.6.1.1 Socio-economic benefits

It is claimed that organic farming contributes to increased employment in rural areas, due to organic farms generally being smaller and more diverse and dynamic in terms of routes to market (e.g. vegetable box schemes) and the encouragement of on-farm processing and direct marketing.

Despite the comprehensiveness of covering a wide range of environmental, welfare and ethical criteria and whole chain control and monitoring, organic food is probably one of the production and labelling systems least associated with either environment protection or ethical operation. Because the range of criteria that organic farming delivers is so broad, it is too difficult and complex to convey to consumers with a simple "logo" or "message." Most consumers buy organic foods primarily for personal benefits, such as avoidance of pesticides, better flavour or less additives. Welfare and environment do feature as reasons for purchase, but these are usually lower down the list of reasons for purchasing.

The full requirements of organic standards are too wide-ranging to go into detail in this chapter, so this will give an overview of the finished (consumer) food product standards. Most regulatory organic standards, including EU regulations, have two categories of organic content. For products to be labelled as organic in the product title or description, a minimum 95% organic ingredients of agricultural origin (i.e. excluding ingredients such as salt, water and additives) is required. For composite products with less than 95% organic agricultural ingredient content, the product can be referred to either as "made with organic" or just make reference to organic ingredients in the ingredient declaration.

In the EU, all retail products must now have a common European organic logo clearly displayed, although other national and/or private logos of organic associations and certification bodies can be displayed in addition to the EU logo (Figure 9.1).

This range of differing organic logos can be confusing to the consumer who may not always understand what they mean. The only worldwide organic logo is the Demeter Biodynamic logo (www.demeter.net/) (accessed 5 June 2013) (Figure 9.2).

Figure 9.1 European organic logo.

Figure 9.2 Demeter organic logo.

Biodynamic agriculture is an enhanced form of organic agriculture in which various agricultural activities are undertaken according to "cosmic constellations" and with the help of various "preparations." It is described as a "holistic" approach and is based on the original ideas of Rudolph Steiner. Use of preparations made from healing plants, minerals (quartz crystal) and cow manure, sprayed onto the land or inserted into compost, are used to enhance root growth and soil life as well as to fortify the plants. Products produced to Demeter standards meet all EU organic regulatory standards, and those of most other organic regulatory regimes.

Demeter is the only ecological association that has built up a network of individual certification organisations worldwide. Demeter International has 17 members and seven guest-members from Europe, America, Africa and New Zealand. Thus Demeter International represents c.5.000 Demeter farms, with nearly 150.000 hectares in more than 50 countries.

9.6.2 Fairtrade

The Fairtrade (www.fairtrade.net) mark is a worldwide recognised product logo and the scheme aims to offer farmers the following:

- a guaranteed minimum price for their product (unique to the Fairtrade system),
- a social premium on top of the minimum price to spend on improving their community,
- access to credit to improve their business and
- help to build sustainable trade relationships

The Fairtrade vision is www.fairtrade.org.uk/what_is_fairtrade/fairtrade_foundation.aspx (accessed 15 June 2013) (Figure 9.3):

"A world in which all producers can enjoy secure, sustainable livelihoods, fulfil their potential and decide on their future."

The role of the Fairtrade Foundation is to work with businesses, community groups and individuals to improve the trading position of producer organisations in the global south to deliver sustainable livelihoods for farmers, workers and communities.

Figure 9.3 Fairtrade Mark.

Various worldwide standards have been developed based on two main ethical categories:

- Producer Standards:
  - Social Development (health and safety, worker rights)
  - Economic Development
  - Environmental Development (pesticides, sustainability)
- Trader Standards:
  - Fairtrade Minimum Price and Fairtrade Premium
  - Pre-finance when requested
  - Long-term commitment and planning

Fairtrade International is a non-profit global movement known as Fairtrade Labelling Organisation (FLO). This association sets standards, agrees prices and premiums, develops trade and supports producers who are largely based in the southern hemisphere. It is supported by labelling initiatives which are national organisations currently made up of 19 organisations covering 24 countries, mostly "consumer" countries based in the northern hemisphere and Australasia.

According to Fairtrade (www.fairtrade.org.uk/includes/documents/cm_docs/2012/2/2011 Annual Report and Accounts – FINAL FULLY SIGNED.pdf) (accessed 15 June 2013), around 7.5 million people (farmers, families, workers and communities) benefit from the premiums paid. There are 991 certified producer organisations and Fairtrade products are sold in 120 countries (at 2011). Fairtrade labelling covers not only a range of commodity food products such as cocoa, coffee, tea, rice, fruit and sugar, but also some non-food products such as cotton, flowers and gold. Fairtrade state that their logo enjoys "awareness" levels of 77%, with consumer trust levels at 90%.

The Fairtrade standards are a voluntary scheme and not supported by any legal backing, other than contractual arrangements. The standards and certification scheme are subject to ISEAL review.

The standards are wide-ranging and cover: primary production; secondary products; trade standards; national standards (covering handling, processing and labelling); and product category standards.

The standards are supported by Price and Premium Tables, which specify minimum pricing for commodities and Fairtrade premiums. Standards cover contractual and trading relationships, handling (such as identification, documentation and separation), use of the Fairtrade Mark (logo) and record keeping of purchases, sales and transactions.

One of the lesser known aspects of Fairtrade is a concept of "documentary traceability" for the following world-traded bulk commodities: coffee, sugar, fruit juice and cocoa. These products, when packed as Fairtrade may not have been purchased as Fairtrade. However, providing the operator can demonstrate that sufficient Fairtrade raw material has been purchased to cover total sales, and then this is acceptable. For all other food products, there must be full "physical traceability" in place, meaning that ingredients stated as Fairtrade must be traceable back to a Fairtrade source on a batch-by-batch basis. There are also compositional standards for composite (multi-ingredient) products.

Figuire 9.4 Rainforest Green Frog Logo.

9.6.3 Rainforest Alliance

Rainforest Alliance (www.rainforest-alliance.org.uk) is a combined environmental and ethical initiative based on the three pillars of sustainable agriculture, social equity and economic viability. It was established in 1992 under the auspices of the Sustainable Agriculture Network (SAN), a coalition of NGOs to work with farmers to protect wild life, wild lands, worker's rights and local communities, particularly in areas where native forest is under pressure from agriculture. Agricultural expansion is responsible for 70% of global deforestation, and is the single greatest threat to tropical forests. It now operates in 25 tropical countries, and more than 2 million farmers are claimed to benefit from Rainforest Alliance certification (Figure 9.4).

Rainforest Alliance uses the SAN standards as a basis of certification. These use the following ten core principles as a basis of standard setting:

(http://sanstandards.org/sitio/subsections/display/7) (accessed 4 June 2013):

1. a management system – social and environmental management systems
2. ecosystem conservation – this includes prohibiting logging, other deforestation and maintaining natural vegetation barriers.
3. wildlife protection – farmers should take specific steps to protect endangered species.
4. water conservation – practices to minimise water use and to avoid water contamination
5. working conditions – compliance with International Labour Organisation conventions
6. occupational health – training of staff to create safe working environments
7. community relations – farms to inform and consult communities about activities
8. integrated crop management – monitoring of pest and disease to minimise chemical use
9. soil conservation – prevention of erosion, by use of organic matter
10. integrated waste management – managing waste through reduction and recycling.

The certification is based on the SAN Sustainable Agriculture Standard http://sanstandards.org/sitio/subsections/display/9 (accessed 30 November 2013).

The objective of the standard is to encourage farms to analyse, and consequently mitigate, environmental and social risks caused by agriculture activities through a process that motivates continual improvement. The standard is based on the themes of environmental soundness, social equity and economic viability. Compliance is based on a scoring system.

General Compliance: In order to obtain and maintain certification, farms must comply with at least 50% of the applicable criteria of each principle and at least 80% of the total applicable criteria of the Sustainable Agriculture Standard.

The producer organisations are fully inspected and certified, with a "chain of custody" system for the supply chain downstream. Certification is either direct by Rainforest Alliance certification (RA-Cert), or by contracted auditing bodies, although the farms still become "Rainforest certified." The main product categories are tea, coffee, cocoa, fruit and vegetables from Africa, Latin America, Asia and Hawaii.

The "green frog seal" is seen increasingly on retail products and products used in food service outlets, the first being in 2004 when Kraft and Proctor & Gamble first launched Rainforest Alliance certified coffee in retail stores. The seal can now be found on a wide range of products and outlets, including Lipton Tea, McDonalds coffee, Wholefoods Market, Unilever and Chiquita.

9.6.4 UTZ

UTZ (www.utzcertified.org) is a more product-specific environmental and ethical initiative. It was founded in the 1990s by a Dutch coffee roaster and a Belgium-Guatemalan coffee grower after observing the limitations of the coffee market and consumer labels available at the time. The word "UTZ" was taken from the Mayan "Utz Kapeh" meaning "good coffee." The certification scheme was launched in 2002 as a sustainable coffee programme, although it has expanded to include cocoa and tea in subsequent years. The UTZ vision is stated thus (www.utzcertified.org/en/aboututzcertified/vision-mission) (accessed 4 June 2013):

- a world where sustainable farming is the norm is a world where: farmers implement good agricultural practices and manage their farms profitably with respect for people and planet,
- industry invests in and rewards sustainable production, and
- consumers can enjoy and trust the products they buy.

The UTZ programme is a producer-based certification scheme with chain of custody certification of downstream handling and processing.

UTZ Certified developed product-specific codes of conduct for coffee, cocoa, tea and rooibos, as well as chain of custody standards and a certification system. To achieve UTZ certification, producers must comply with the economic, social and environmental criteria set out in the codes of conduct. This focus aims to improve the income of farmers and their communities mainly by sustainable economic improvement by improving farming methods, which in turn increases yields and quality of the product, attracting a better price on the market place, rather than relying solely on artificial price support or the vagaries of commodity traders. UTZ want the producers to offer sustainable ingredients to their clients, and the logo is not promoted among consumers, since UTZ is not a brand but a programme of production. UTZ want consumers to buy sustainable products by following the brands they like, which happen to use ingredients produced to UTZ standards (Figure 9.5).

The UTZ standards allow the use of agrochemicals where required, but aim to encourage minimal and responsible use through good farming practice and recycling

Figuire 9.5 UTZ logo.

of nutrients. In terms of market penetration, the UTZ Certified Annual Report 2012 (www.utzcertified.org/images/stories/site/pdf/downloads/utz annual report 2012-final.pdf) (accessed 4 June 2013) states that 13% of all world cocoa production, 8% of the global coffee harvest and 2% of all tea were UTZ certified. The volume of UTZ certified coffee experienced 50% growth from 2011/12 to 2012/13. Cocoa enjoyed a 178% growth in the same period, albeit from a small base to currently 534 614 tonnes of cocoa. UTZ has the support of a range of major European and worldwide brands such as Ikea, Lidl, REWE Group, Tchibo, Nestle, Mars and Migros. This demonstrates increasing consumer support and interest in such schemes.

9.6.5 Marine stewardship council

Ethical and environmental standards are not restricted to agriculture; the over fishing of the oceans has been an ever-increasing problem, due to increased demand and massively improved technology for searching and catching of fish. This has led to a five-fold increase in the landing of fish in the last 50 years, and a quarter of fisheries are classed as depleted or overfished. The Marine Stewardship Council (MSC) (www.msc.org) was formed in 1997, a result of two global organisations, WWF and Unilever, wanting to tackle the issue of seafood sustainability, based on concerns following the collapse of the Newfoundland cod fishery in 1992. Its main aims were to recognise and reward fisheries that fish sustainably, work with fishery and commercial partners to build a market for sustainable seafood and give buyers and consumers an easy way to find seafood from a sustainable fishery – the blue Marine Stewardship Council (MSC) ecolabel.

The MSC operates a "third party" certification programme. Certificates are issued by certifiers who are independently accredited to be able to perform assessments of fisheries and businesses under licence to the MSC standards. MSC certifiers are accredited and the MSC scheme follows codes set by ISEAL. There are currently over 200 fisheries in the MSC programme. A "fishery" in the MSC programme may include one or more "units of certification."

- A unit of certification is usually defined by reference to the following:
- Target fish species and stock
- Geographic area of fishing
- Fishing method, gear, practice and/or vessel type

MSC is one of the more challenging schemes to monitor; however, they use "chain of custody" certification to trace products through the supply chain from capture to retail sale. In 2007 they launched a Traceability and Assurance in the Supply Chain (TASC) project to make their programme even more robust. This includes trace-backs down the supply chain from the point of sale to the fishery. It also includes DNA testing, random checks by certifiers and a system for double-checking claims. In terms of monitoring the impact of the scheme, MSC employed researchers in 2005 to undertake an initial evaluation of the environmental impacts of MSC certification on the management practices of ten fisheries. The study looked at the period after certification. The researchers investigated more than 60 separate actions that ten fisheries undertook between 2000 and 2005 after they had been certified for at least a year. The results were compiled in a report (Agnew *et al.*, 2006) and concluded that:

- instances of environmental gain can be detected in fisheries that have been certified,
- instances of environmental gain could be attributed directly to the scheme,
- however, some instances of environmental gain were attributed to elements that were not related to certification conditions. This point highlights the difficulties of demonstrating and monitoring environmental benefits in natural ecosystems.

9.6.6 Pro terra

The ProTerra (http://www.proterrafoundation.org/) Certification Program is a relatively recent scheme aimed primarily at production and certification of non-GMO food and feed ingredients, and gives assurance that the certified organisation complies with strict standards surrounding issues such as:

- Protection of the Amazon and other High Conservation Value Areas,
- Good labour practices including workplace safety, equal opportunity, protection of children and forbidding slave labour,
- Protection of the community, indigenous people and small holder farmers,
- Protection of traditional land use,
- Agronomic sustainability (minimisation of soil erosion, pesticide use and water usage),
- Rigorous non-GMO requirements (< 0.1% adventitious GMO).

The standard covers three main categories: primary production, storage and distribution and finished product certification. Finished products must use at least 70% ProTerra certified ingredients to carry the ProTerra mark. To ensure traceability and deliver consumer trust, a ProTerra Traceability Database has been developed to ensure full traceability of every lot and consignment of ProTerra-certified material at every point in the chain of custody back to the original group of farmers that produced the certified raw product (Table 9.2).

9.7 A review of industry labels and schemes

In addition to the international and largely independently certified ethical and environmental labelling schemes, there are a plethora of industry and product-specific

Table 9.2 Comparison of main schemes

Scheme	Organic	Fairtrade	Rainforest Alliance	UTZ certified	MSC	Proterra
Regulated	In 49 countries	No	No	No	No	No
Accredited certification	ISO65/EN45011		ISO65 for certification	ISO65/EN45011	ISO65/EN45011	ISO65/EN45011
Accredited standards		ISEAL compliant	ISEAL compliant	ISEAL compliant	ISEAL compliant	
Whole chain certification	Yes	Yes	Certification and chain of custody	Certification and chain of custody	Fishery certification and chain of custody	Yes
Environmental standards	Yes	Yes	Yes	Yes	Yes	Yes
Animal welfare standards	Yes	No	Cattle only	No	No	No
Ethical/social standards	In IFOAM principles	Yes	Yes	Yes	Yes	Yes
Synthetic agro-chemicals	Highly restricted	Minimal and responsible use	SAN list of prohibited pesticides	Responsible use	N/A	Requires minimal use
Genetic modification	Prohibited	Prohibited	Prohibited	Not stated	N/A	Prohibited
World sales value	$59.1 bn (2012)	€4.9 bn (2011)	Not stated		$3.5 bn 2012–13	Not stated
World area of production	37.2 m ha (at 2012)	Not stated	75 m ha forest and 1.53 m ha farmland (2012)		8% of wild capture fisheries (2013)	Not stated

Source: Willer and Kilcher, 2002

schemes. They are too numerous to address in detail in this chapter, but a selection are described to give an indication of their range.

9.7.1 Marks & Spencer Plan A

Marks and Spencer launched Plan A (http://plana.marksandspencer.com) in January 2007, setting out 100 commitments to achieve in 5 years. They have now extended Plan A to 180 commitments to achieve by 2015, with the ultimate goal of becoming the world's most sustainable major retailer. Plan A is not just a product "logo," but is an organisational plan to ensure the long-term sustainability of the business and it is called "Plan A," as they say there is no "Plan B." The plan has seven pillars:

- Make Plan A "how we do business",
- Involve customers,
- Climate change – reducing emissions and energy usage from production to retail,
- Waste – zero waste to landfill from stores and reduction of packaging and food waste,
- Natural resources – sourcing food ingredients from sustainable sources, although in 2013 they removed their requirement for non-genetically modified (GM) feedstuffs for livestock,
- Fair partner – paying a fair price to suppliers and supporting local communities,
- Health and wellbeing – working with UK Government's health responsibility deal.

They produce an annual report of progress with the plan, which in 2012 stated they had so far met 139 of their 180 commitments (http://planareport.marksandspencer.com/review-of-the-year) (accessed 15 June 2013).

9.7.2 Hofer Supermarket Group, Austria "Zuruck zum Ursprung"

The scheme "zuruck zum ursprung" (www.zurueckzumursprung.at) or "back to source" has been used to launch and promote the organic brand of the Austrian supermarket "Hofer" which is part of Aldi Group, Germany. The "ground values" are:

> *Regionality; environmental protection; non-GMO; high animal welfare; food quality; transparency; fairness to the consumer and fairness to the farmer.*

The strong point of this scheme is that they have managed to encapsulate the benefits of organic farming into one simple logo indicating reduced CO_2 and water usage, and being better for nature and biodiversity. The website lists all of the products within the range and then gives a breakdown how purchasing the "back to source" labelled product is better than purchasing a similar non-labelled product in terms of CO_2 output and water usage. The scheme is monitored independently and there is a consumer portal which facilitates tracing products that have been purchased back to source by batch code.

9.7.3 REWE group pro-planet

This logo has been developed by the REWE (www.proplanet-label.com/) retail group of Germany and the scheme aims to be a "navigation system for sustainable products." It covers both food and non-food products and these principles have a positive effect on:

- sustainability,
- ecology,
- social,
- economic properties of the products that are labelled.

It analyses the "value added chain" and finds solutions to ensure the minimal impact of products on the environment and people during the product life-cycle of production, processing and consumption. The scheme has been developed and is overseen by an independent advisory board formed from NGOs and state institutions.

9.7.4 Starbucks coffee and farmer equity (CAFE)

This is a scheme introduced by Starbucks (www.starbucks.com/responsibility/sourcing/coffee) in cooperation with Conservation International to develop and apply a comprehensive set of environmental, social and economic guidelines toe to source ethical coffee. These guidelines are known as CAFE Practices. Starbucks use these to evaluate and reward producers of high-quality, socially responsible and sustainably grown coffee. By 2015, Starbucks aim for all of their coffee to be third party verified or certified, either through CAFE Practices, Fairtrade, or other externally audited systems.

9.7.5 Roundtable on sustainable palm oil

RSPO – www.rspo.org certification is a seal of approval that the palm oil used in the product is produced sustainably and volumes are traceable. It is estimated that palm oil and derivatives are present in one in ten food products on supermarket shelves. Producers are certified through strict verification of the production process by accredited third party certifying agencies. The certified sustainable palm oil is traceable through the supply chain by certification of each facility along the supply chain that processes or uses the certified oil.

9.7.6 Dolphin friendly tuna

There are a number of "dolphin friendly" schemes; however, one of the leading ones is operated by Earth Island Institute (www.earthisland.org/dolphinSafeTuna/consumer) and is a US-based not-for-profit organisation that monitors tuna companies around the world to ensure the tuna are caught by methods that do not harm dolphins and that protect the marine ecosystem. Their standards aim to prevent harm to dolphins, and they claim they are adhered to by more than 90% of the world's tuna companies. The Which report 2010 referenced earlier in this chapter noted that the dolphin friendly logo is recognised by 44% of consumers, whereas only 15% recognised the MSC logo.

It has the following standards:

1. No intentional chasing, netting or encirclement of dolphins during an entire fishing trip;
2. No use of drift gill nets to catch tuna;
3. No accidental killing or serious injury to any dolphins during net sets;
4. No mixing of dolphin-safe and dolphin-deadly tuna in individual boat wells (for accidental kill of dolphins), or in processing or storage facilities; and
5. Each trip in the Eastern Tropical Pacific Ocean (ETP) by vessels 400 gross tons and above must have an independent observer on board.

The website lists "dolphin friendly" and "dolphin deadly" tuna companies and publishes an annual report of monitoring activities.

9.8 Economics of the main labelling schemes

Like all things, there is a cost to the management, certification and monitoring of these schemes. How the burden of costs and benefits is distributed across the supply chains varies from scheme to scheme, but the simple answer is that the consumer ultimately pays through purchasing the labelled products. How the price paid flows down to the producer and funds the necessary certifications and monitoring varies between the schemes. The funding of the main schemes is briefly described below.

9.8.1 Organics

The economics of organics is primarily financed through organic products attracting a price premium; however, this is strictly market dictated rather than set by any particular body. The price premiums are not just an "organic premium" but reflect the additional costs of raw materials, certification costs of operators, and the often "hidden" costs of shorter production runs, higher waste costs proportional to volumes processed and higher packaging costs due to artwork/set up costs relative to quantities processed. Producers may benefit from higher prices, but this is not guaranteed. In 2012, the European Commission conducted a survey on the cost of certification to the organic industry. (Anon, 2012). Although this report was based on organic certification, its findings could be applied to a range of similar certification schemes.

9.8.2 Fairtrade

This scheme has a minimum pricing structure for the purchase of a wide range of products and commodities that specifies minimum pricing payable to producer organisations. Additional to this, a producer premium is payable, direct to producer cooperatives or organisations. For processing and traders in the consumer countries, a percentage fee of 1.7% is payable on the value of products sold.

9.8.3 Rainforest Alliance

In addition to funding from private foundations, government grants and individual donors, the Rainforest Alliance and SAN implemented a participation fee for companies on 1 October 2010. The fee is levied on the volume of Rainforest Alliance certified goods traded in the marketplace. Coffee importers are responsible for paying the participation fee, not the farmers. Farmers pay for technicians and auditors, although visits are usually performed by local organisations, keeping costs lower. As per organic certification systems, smallholders may organise an internal management system and seek certification as a group, thus reducing auditing and administrative costs.

9.8.4 UTZ certified

This scheme is funded by the purchasers and traders of products with support of EU funding and NGO funding. The farmers receive a premium for their products based on quality and increased yield and do not have to pay for inspection and certification.

9.8.5 MSC

A flat fee is in place for all MSC licence holders ranging from $250 to $2000 based on sales value of MSC certified non-consumer and consumer-facing products. Additional to this is a fee ranging from 0.5% down to 0.3% levied on the wholesale price of consumer-facing products (e.g. retail packed or those on retail display).

9.9 Summary

The range of schemes and labels that now make some form of environmental and ethical claim are bewildering. The simpler ones are easy to understand by the consumer – e.g. "Fairtrade" or "Rainforest Alliance." However, schemes such as the organic scheme, which arguably deliver a wider range of benefits and are operated within a legal framework, are less well understood and the benefits are harder to communicate because of the range and complexity of issues that the system addresses. Newer schemes, such as Rainforest Alliance and UTZ Certified, have experienced significant growth in sales over recent years. Much of this is due to adoption by multi-national brands as part of their Corporate Social Responsibility policies rather than by specific consumer choice. The consumer tends to choose well-known brands which happen to be Rainforest Alliance or UTZ certified, whereas products with Fairtrade or organic labelling are purchased because they are labelled as such.

In order for labelling schemes to ensure consumer "buy in," trust, understanding and ultimately support through purchase choice, a scheme must consider the follow points:

- A relevant cause that is important to consumers
- Products that have consumer appeal and meet consumer expectations
- Clear, consistent and transparent standards/requirements

- Certification is independent of the logo users (third party certification)
- Mechanisms for review and improvement
- Independently verifiable
- Products that are easily traceable back to source
- A logo that has good customer (household, institutional and corporate) recognition and demand

This chapter has highlighted the complexity, breadth and depth of this subject area. The majority of these schemes aim to deliver/influence a purchasing choice, in a retail shopping environment where "ethical and environmental" considerations often play second fiddle to more immediate purchasing pressures such as time, cost and social pressures ("the trolley dash"), the urge is often to get the shopping done as quickly and painlessly as possible.

So, how do shoppers find their way through the maze of labels? Modern technology helps. A number of the scheme websites now have short video clips summarising the benefits of the scheme. Question and Answer sheets can deliver key messages and use of quick response (QR) codes can deliver complex messages quickly and conveniently at the point of purchase via smart phones. On-line shopping is also gaining in popularity so consumers can quickly search for on-line information at point of purchase. One very useful resource is the ecolabel index (http://www.ecolabelindex.com/) which tracks over 400 "eco-labels" of which 146 are related to food. It summarises what each label stands for, how it is monitored (e.g. independently certified) and what its managing structure is (e.g. whether a "for profit" organisation or "not-for-profit"). For consumers who are more committed to making ethical and environmental purchasing decisions, then the only real way to make appropriately informed decisions is to do some good, old fashioned research. However, this must be done away from the shopping environment, and consumers would need to make a concerted effort to learn which labels they wish to support so they can look for them while shopping. It must also be borne in mind that an ever-increasing food spend takes place outside of the home in terms of eating out, where an approximate 40% of weekly food spend now lies.

9.10 Future trends

With such a wide array of schemes and labels, consolidation and simplification are inevitable. Corporate social responsibility pressures have led a number of companies to develop their own schemes as discussed above. These have the advantage that consumers already trust the brand, although they are less independent than the third party schemes. The other concern is the added cost of monitoring and running the schemes, which ultimately end up being paid for by the consumer.

For the future, some regulatory control may be a necessity or use of "umbrella" logos that incorporate a number of environmental and ethical schemes/benefits under one "super label," but again this adds another layer of cost and complexity. An alternative is some form of "scoring system" which gives a rating on the effectiveness, efficacy and integrity of eco and ethical label schemes, based on internationally defined standards, similar to nutritional "traffic light" labelling in the UK.

For locally produced foods, organisations such as the UK-based Wholesome Food Association (www.wholesome-food.org/) is a peer-to-peer scheme which is self-monitoring via an "open gate" policy, thus avoiding external monitoring costs. A number of schemes are reducing monitoring costs by use of "internal control systems" where producers monitor themselves and certifiers audit the application of the control system and sample a small number of growers.

However, as the world population continues to grow, and non-renewable resources become depleted, there will come a time when all food will have to be produced sustainably and with minimal impact on the environment, while being ethically and socially just, regardless of how it is labelled.

References

Agnew, D., Grieve, C., Orr, P., Parkes, G. and Barker, N. (2006), *Environmental Benefits Resulting from Certification Against MSC's Principles and Criteria for Sustainable Fishing*. MRAG UK Ltd and Marine Stewardship Council, London. p. 134.

Anon (2012), *Economic Analysis of Certification Systems for Organic Food and Farming*. European Commission (www.certcost.org) (accessed 15.06.13).

European Commission (2010), *Special Eurobarometer Report 354, Food-Related Risks*. European Food Safety Authority, Belgium.

European Regulation (EU) no 1169/2011 of the European Parliament and of the Council of 25 October 2011, on the provision of food information to consumers. *Official Journal of the European Union*, L304/44.

FAO (2011), *Organic Agriculture and Climate Change Mitigation, a Report of the Round Table on Organic Agriculture and Climate Change*, Food and Agriculture Organisation of the United Nations, Rome.

Food Ethics Council (2009), *Ethical Consumption, Solution or Problem?* A report of the Business Forum Meeting of 20th January 2009. Food Ethics Council.

Foster, C., Green, K., Bleda, M., Dewick, P., Evans, B., Flynn, A. and Mylan, J. (2006), *Environmental Impacts of Food Production and Consumption: A report to the Department for Environment Food and Rural Affairs*. Manchester Business School. DEFRA, London.

GEN (2004), *Introduction to Ecolabelling*, Global Ecolabelling Network information paper. p.1, Ottawa, Ontario, Canada. www.gen.gr.jp.

ISO (2012), *Environmental Labels and Declarations. How ISO Standards Help*. ISO Central Secretariat. Geneva, Switzerland. http://www.iso.org/iso/home/news_index/news_archive/news.htm?refid=Ref1705.

Soil Association (2012), *Organic Market Report*, Soil Association, Bristol. p.7.

Venkat, K. (2012), Comparison of twelve organic and conventional farming systems: a life cycle greenhouse gas emissions perspective, *Journal of Sustainable Agriculture*, **36**:6, 620–649.

Which (2010), *Making Sustainable Food Choices Easier. A Consumer Focussed Approach to Food Labels*. www.which.co.uk/documents/pdf/making-sustainable-food-choices-easier-which-report-231317.pdf.

Willer, H. and Kilcher, L. (Eds.) (2012), The World of Organic Agriculture – Statistics and Emerging Trends 2012. Research Institute of Organic Agriculture (FiBL), Frick, and International Federation of Organic Agriculture Movements (IFOAM), Bonn V 3.0 23/02/2012.

Further reading

There is an almost inexhaustible range of articles, reports and reviews about environmental and ethical labelling schemes. The author suggests visiting the websites of the various schemes above but also particularly recommends:

www.ecolabelindex.com
www.isealalliance.org
www.foodethicscouncil.org

Labelling of genetically modified (GM) ingredients in foods and beverages

10

B. Gallani
Food and Drink Federation (FDF), London, UK

10.1 Introduction

Genetic modification (GM) is the alteration of genetic material (i.e. DNA) through the use of a range of laboratory techniques. It has been enabling plant scientists to transfer genetic material between species that cannot be crossed using conventional breeding and has resulted in the introduction of desirable traits into crops. Genetic modification and conventional breeding are seen as complementary approaches aimed at offering potential benefits to farmers as well as consumers. In Europe, the processes for the evaluation and authorisation of GM foods are laid down in Regulation (EC) No 1829/2003 on GM food and feed, which came into force in April 2004. The safety assessment for each of the submitted dossiers is carried out by the European Food Safety Authority (EFSA), according to its published guidelines for the assessment of GM plants. Regulation (EC) No 1830/2003 on traceability and labelling of GM food and feed was developed in recognition of the fact that, independently of a positive safety assessment carried out by EFSA, some people might choose not to buy or eat GM foods. Although very few GM crops are licensed for cultivation in Europe, many crops have been authorised for cultivation in other parts of the world. Given the global trade of raw materials between different continents, it is important that international guidelines are introduced with regard to unavoidable low levels – known as low-level presence (LLP) – of GM products that are authorised for use in a given country but are not yet approved in an importing country.

10.2 Genetic modification (GM) in food production: an overview

The GM Timeline published on the Food Standards Agency (FSA) website, and capturing the history of humans' manipulation of genes in plants and animals, starts in 10,000–9000 BC, when people began to plant crops rather than relying on hunting and gathering for food. Selective breeding by Peruvians in 3000–2000 BC was aimed

Advances in Food and Beverage Labelling. http://dx.doi.org/10.1533/9781782420934.3.177

at identifying potatoes from around 160 wild species with lowest levels of poisons and growing them for food. Assyrians and Babylonians in 1000–700 BC were already pollinating date palms by hand, but it was not until the late seventeenth century AD that cells and plant structures and functions were described in detail, together with the first systematic breeding of flowers. In 1866 Gregor Mendel published his work on pea plants, establishing a pattern of inherited characteristics. The twentieth century saw the development of a number of techniques to better see structures inside the cells, including chromosomes. Thomas Hunt Morgan's team at Columbia University set the standard for gene mapping, through their work on the fruit fly carried out between 1907 and 1917. At the beginning of the last century, a number of scientists researched the mutations caused by radiation and chemicals on both animals and plants. In 1953 Francis Crick and James Watson revealed the double helix structure of the DNA, paving the way to a better understanding of how genes work. In 1980s, scientists developed the GM crop plants that are now grown in the fields. China was the first country where GM crops (a virus-resistant tobacco and a tomato) were put on sale. The start of widespread use of GM crop plants in the US was in 1994, with the introduction of a GM tomato variety, followed by insect-resistant corn in 1995 and herbicide-resistant soy beans in 1996. In 2004 Golden rice, a variety that contains beta-carotene, which is converted in the body into vitamin A, started to be grown in some parts of the world, followed in 2005 by salt-tolerant tomatoes and sunflowers resistant to white mould. Although not a comprehensive list, the FSA GM Timeline provides the context for the current, lively debate on GM.

A briefing note published by the Science Media Centre (SMC) (2013) states that no adverse health impacts have yet been documented from the consumption of GM crops. In 2011, the total global area sown with GM crops was estimated as 160 million hectares in 29 countries, the equivalent of three times the area of Spain and approximately 10% of the world's food crops. Two varieties of GM crops are licensed for cultivation in the EU: a type of maize and a potato variety. No GM crops have yet been approved for commercial growing in the UK, although research on potato and wheat has recently resumed. GM crops can be imported and used to produce ingredients for human consumption and for animal feed, with currently 44 varieties of GM plant which can be imported into the EU and UK.

Modern technologies, including GM, offer considerable potential to improve the quality and quantity of the food supply and could contribute to sustainability by helping to produce more food using fewer resources and with less impact on the environment. It is broadly accepted that the impact of this technology must be objectively assessed through independent and thorough scientific investigation, and that robust controls are necessary to protect the consumer and the environment from potential risks.

The Foresight Future of Food and Farming report (2011) concluded that to feed a growing global population, the same amount of land will have to produce more food. This is described as "sustainable intensification" and, despite polarised views on the way food production is carried out, no single approach is capable of delivering "sustainable, resilient high levels of productivity and value." This means that conventional plant breeding, improved agricultural practices and biotechnology could be used to increase yields from the same area of land while using the same inputs, such as fertiliser and water.

The Department for Environment, Food and Rural Affairs (Defra) recognises that GM technology could deliver benefits provided it is used safely and responsibly, as one of a range of tools to address the longer term challenges of global food security, climate change and the need for more sustainable agricultural production. Defra also supports the implementation of pragmatic and proportionate measures to segregate GM crops from conventional and organic crops, so that choice can be exercised and economic interests appropriately protected. The Defra position on GM also states that developing countries should have fair access to such technology and make their own informed decisions regarding its use.

10.3 European approval processes

The protection of human health and the environment is undoubtedly the main priority when it comes to the authorisation of GM crops. In Europe, the approval for planting GM crops or marketing GM food or feed products is subject to a robust risk assessment to demonstrate that the product is safe for consumers and the environment. GM product applications are assessed on a case-by-case basis, taking into account all the available scientific evidence, including detailed considerations of the potential for toxic, nutritional and allergenic effects. In addition, GM foods are authorised for sale only if they are considered not to mislead consumers and not to be of less nutritional value than the foods they are intended to replace.

The role of the EFSA, defined in Regulation (EC) No 1829/2003 and Directive 2001/18/EC, is to assess any possible risks of genetically modified organisms (GMO) to human and animal health, and the environment. Ultimately, it is the role of the European Commission and Member States, in their capacity as risk managers, to authorise GMOs, while EFSA's role is strictly limited to providing scientific advice.

EFSA's assessments are carried out by the independent scientific experts that make up the EFSA GMO Panel, with the support of a number of specialised Working Groups of external experts in fields such as allergenicity, ecology, microbiology, toxicology, plant physiology and molecular genetics. EFSA has published seven main guidance documents for the risk assessment of GMOs and derived food and feed, and several other supporting documents describing the data that applicants must include in their application dossiers, which include all the necessary studies on human and animal safety and on environmental impact.

When considering each application, the EFSA Panel takes into consideration the following aspects of each GM product: its molecular characterisation; its compositional, nutritional and agronomic characteristics; its potential toxicity and allergenicity; and its potential environmental impact.

A comprehensive environmental risk assessment (ERA) is intended to identify any possible adverse effects and, according to the legal framework, applications for GM plants intended for cultivation must include an initial ERA that has to be carried out by one Member State.

Under EU legislation, a Member State can invoke a specific safeguard clause to provisionally prohibit the cultivation or use of a GMO in its territory based on scientifically based concerns regarding the safety of the GMO. To support the legal decision-making process, the Commission may ask EFSA to provide a scientific opinion on the information presented by Member States. The final decision on each individual authorisation is taken by Member States through a vote in the Standing Committee on the Food Chain and Animal Health (SCoFCAH). The European Commission maintains a Register of GM food and feed materials that are authorised in the EU. In UK, the FSA leads the safety of GM food and feed and on applications to market GM food and feed products. The Department for Environment, Food and Rural Affairs (Defra) leads the environmental safety of GMOs and wider issues surrounding the use of GM crop technology.

10.4 European regulation of GM food and feed

In order to ensure the safety of GM technologies, the European Union has established a robust legal framework. The main objectives of the regulation on GM food and feed are to protect human and animal health by introducing an independent assessment before any GM food and feed is placed on the market; to have in place harmonised procedures for risk assessment and authorisation of GM food and feed that are efficient, time-limited and transparent; to ensure clear labelling of GM food and feed; and to set labelling requirements for GM feed which provide farmers with accurate information on the composition and properties of the feed they buy.

Regulation (EC) 1829/2003 on GM food and feed has as its main objective consumer safety, while also ensuring that the internal market continues to work effectively. This regulation creates a specific harmonised procedure for the scientific assessment and authorisation of GM food and feed products. It also requires labelling of all GM food and feed which contains or consists of GMOs, is produced from GMOs, or contains ingredients produced from GMOs.

This regulation is supplemented by Regulation (EC) 1830/2003 on the traceability and labelling of GMOs placed on the market.

In addition, Directive 2001/18/EC on the deliberate release of GMOs into the environment outlines the principles for, and regulates, experimental releases and the placing on the market of GMOs in the EU. European guidelines on the co-existence of GM, conventional and organic crops have been developed to prevent contamination of conventional crops by GM crops.

In UK the Environmental Protection Act 1990 (EPA) is the primary legislation that covers the deliberate release of GMOs in England, through the implementation of Directive 2001/18. GM (Deliberate Release) Regulations 2002 supplements the EPA by setting out detailed rules for the implementation of Directive 2001/18, including specific requirements for applications to release GMOs.

Regulations 1829/2003 and 1830/2003 are implemented in England, Northern Ireland, Scotland and Wales through separate documents.

When a GM crop is authorised for commercial use in a country, trace amounts of that crop may become mixed with other varieties of crops, in that country or in transit. As a result, a GM crop that is authorised in an exporting country may be present at low levels in raw materials and finished products that are imported into a country where the GM crop is not authorised. This can cause occurrences of LLP, which are likely to become more frequent as more GM crops are authorised and transported around the world. LLP events lead to disruptions to trade and increased costs to industry and governments, even when unlikely to pose a risk to human or animal health or to the environment.

Many countries have recognised the importance of managing LLP in a proportionate and pragmatic fashion. In Europe, Commission Regulation EU No 619/2011 of 24 June 2011 lays down the methods of sampling and analysis for the official control of feed as regards the presence of GM material for which an authorisation procedure is pending or the authorisation of which has expired. The scope of the regulation covers GM material authorised in a third country, which may be detected in feed at a level up to 0.1% and for which a dossier has been submitted to EFSA. Like feed producers, food producers are dependent on raw material imports and are therefore confronted with very comparable challenges to feed producers. While this regulation represents a first step in an obvious direction, many believe that the scope of the legislation needs to be extended to include food as a matter of urgency.

10.5 GM labelling: principles and regulatory requirements

Independently of the results of a safety assessment, some consumers may wish, for whatever reason, to choose products which do not contain GM-derived ingredients, and food manufacturers have been committed to provide information and choice to their customers.

In the EU, if a food contains or consists of GMOs, or contains ingredients produced from GMOs, this must be indicated on the label. For GM products sold "loose," information must be displayed immediately next to the food to indicate that it is GM (Table 10.1).

The GM Food and Feed Regulation (EC) No 1829/2003 lays down rules to cover all GM food and animal feed, regardless of the presence of any GM material in the final product. This means products such as flour, oils and glucose syrups have to be labelled as GM if they are from a GM source.

The same regulation also provides for a threshold for the adventitious, or accidental, presence of GM material in non-GM food or feed sources. This threshold is set at 0.9% and only applies to GMOs that have an EU authorisation. Any intentional use of GM ingredients at any level must be labelled.

The rules on labelling apply to virtually all foodstuffs: packaged foods, bulk or unpacked foods, and food served in out-of-home settings (e.g. restaurants). In packaged foods the GM ingredient(s) must be qualified as such clearly on the label, for

Table 10.1 Examples of labelling requirements under EC Regulation No 1829/2003 for authorised GMOs

GMO type	Hypothetical examples	Labelling required?
GM plant	Chicory	Yes
GM seed	Maize seeds	Yes
GM food	Maize, soybean, tomato	Yes
Food produced from GMOs	Maize flour, highly refined soya oil, glucose syrup from maize starch	Yes
Food from animals fed GM animal feed	Meat, milk, eggs	No
Food produced with help from a GM enzyme	Cheese, bakery products produced with the help of amylase	No
Food additive/flavouring produced from GMOs	Highly filtered lecithin extracted from GM soybeans used in chocolate	Yes
Feed additive produced from a GMO	Vitamin B2 (Riboflavin)	No
GMM used as a food ingredient	Yeast extract	Yes
Alcoholic beverages which contain a GM ingredient	Wine with GM grapes	Yes
Products containing GM enzymes where the enzyme is acting as an additive or performing a technical function		Yes
GM feed	Maize	Yes
Feed produced from a GMO	Corn gluten feed, soybean meal	Yes
Food containing GM ingredients that are sold in catering establishments		Yes (the FSA's legal view is that labelling is required across EU Member States under EC Regulation 1829/2003).

GM – genetically modified, GMM – genetically modified micro-organism.
Source: FSA's website http://www.food.gov.uk/policy-advice/gm/gm_labelling

example in the list of ingredients, either as "genetically modified…" or as "produced from genetically modified…." Alternatively, this information can be provided, in the same font size, in a footnote to the list of ingredients (Figures 10.1 and 10.2).

In the case of unpacked foods the information "genetically modified…" or "produced from genetically modified…" must be clearly and directly visible as part of the display of the relevant product.

In practice, it is quite difficult to find products in European supermarkets containing GM ingredients.

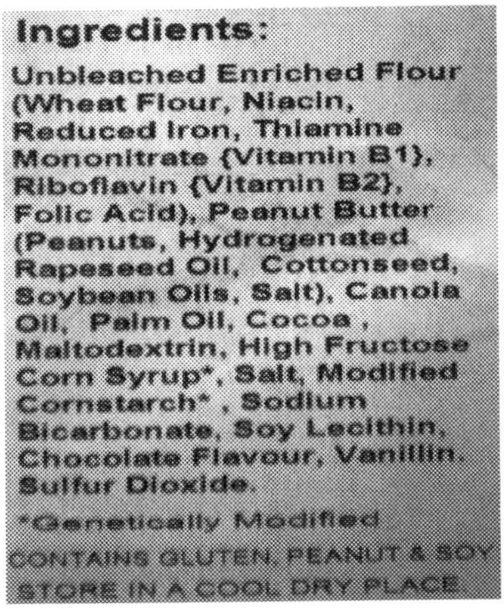

Figure 10.1 Example of GM labelling using an asterisk and footnote clarifying that corn syrup and corn starch in the list of ingredients are "genetically modified."

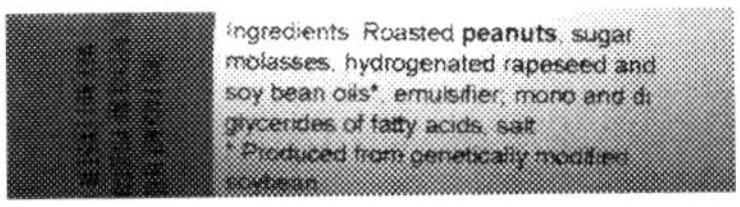

Figure 10.2 Example of GM labelling using an asterisk and footnote to the list of ingredients using the description "produced from genetically modified soybean."

Additives and flavourings manufactured directly from GM organisms all have to be labelled. Some examples are lecithin, vitamins and flavourings from GM soy beans. However, additives and flavourings derived from chemical or other modification of GM raw materials are not usually labelled as GM.

The same applies if GM microorganisms have been used as processing aids and are no longer present in the final ingredient. Enzyme-producing GM microorganisms used in food production include chymosin for making cheese, amylases in bread or glucose syrup production, invertase in sweets and chocolate manufacturing, and pectinases used to make fruit juice or wine. Other GM-related applications that do not require labelling are GM substrates – mainly GM corn, soy and sugar beets used to grow yeast or microorganisms for the production of certain vitamins – and carriers for additives and flavourings. In summary, food or feed produced with GM technology does not fall within the scope of this regulation and therefore does not have to be labelled. This agreement was reached at the SCoFCAH back in 2004. This interpretation was confirmed in the Commission's review of the regulations published in October 2006.

Another product that at present does not require a GM label is honey containing pollen from GM plants, as pollen is not considered an ingredient of honey.

Products such as meat, milk and eggs from animals fed on GM animal feed also do not need to be labelled on the basis that food from animals fed on authorised GM crops is considered to be as safe as food from animals fed on non-GM crops, and on the pragmatic consideration that to date there are no valid methods of detection to identify products of animal origin that have been derived from animals fed on GM feed.

This remains a topical debate despite rigorous independent assessments establishing that GM feed is digested by animals in the same way as conventional feed. In fact, digestion in both animals and humans rapidly degrades the DNA of feed and food constituents and the subsequent uptake of DNA fragments from the intestinal tract into the body is a normal physiological process.

In 2007 EFSA published a statement saying that "a large number of experimental studies with livestock have shown that recombinant DNA fragments or proteins derived from GM plants have not been detected in tissues, fluids or edible products of farm animals like broilers, cattle, pigs or quails." In a subsequent opinion, EFSA noted that "the recombinant sequence is present in the GM plant only as a single or low copy number, which makes the potential absorption a rare event and therefore difficult to detect," and that "when more studies are carried out with more sensitive detection methods, such recombinant DNA fragments may be more frequently found in the future." It is therefore possible that DNA fragments derived from GM plant materials may occasionally be detected in animal tissues, in the same way that DNA fragments derived from non-GM plant materials can be detected in these same tissues although "no technique is currently available to enable a valid and reliable tracing of animal products (meat, milk, eggs) when the producer animals have been fed a diet incorporating GM plants."

The current European regulatory framework does not restrict or prohibit labelling that indicates that specific measures have been implemented to exclude the presence of GMOs in food or feed. A number of "GM-free" labelling schemes, and associated underpinning criteria, have been developed by public or private operators and organisations across Europe. These schemes can be explicit, e.g. clear on-pack messages or logos, or implicit, e.g. embedded in broader product/brand description communicated through media other than the label (Figure 10.3).

In the European Union, GMOs and products derived from GMOs are explicitly excluded from organic production methods. However, the threshold for adventitious and technically unavoidable presence of GMOs in organic products is the same as for conventional products, i.e. 0.9% (Figure 10.4).

In 2012 the Health and Consumer Directorate General (DG SANCO) of the European Commission (EC) launched a study on the state of play of GM-free food labelling in the European Union with the aim of getting a comprehensive overview of existing GM-free labelling schemes in use throughout Europe in order to assess the need for harmonisation. The study followed the publication of a broad report on the evaluation of the EU GM Regulation, which found that there is confusion among consumers because of the proliferation of schemes based on different criteria, therefore making the information questionable in terms of providing consumers with the possibility of making informed choices.

The FSA in the UK has recently carried out new research into public attitudes on GM-free and sought stakeholder views on the potential EU harmonisation of GM-free labelling to inform the on-going discussions in Brussels on this topic.

Figure 10.3 Example of GM-free label on the front of a pack.

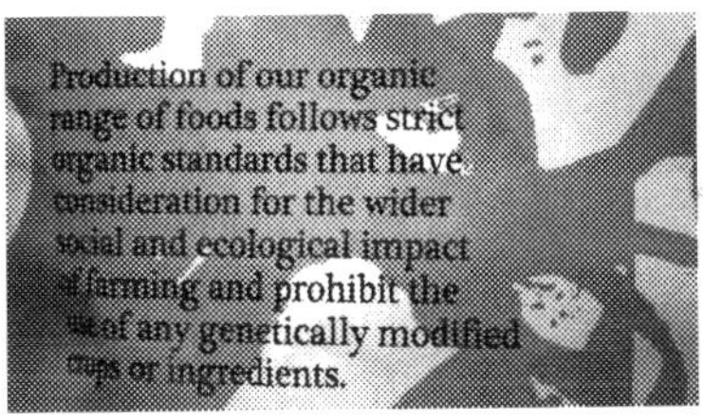

Figure 10.4 Label explaining that organic standards "prohibit the use of any genetically modified food or ingredients."

As predictable, there are mixed views on whether harmonisation in this area is required, especially as rules laid out in Regulation 1829/2003 already cover positive labelling for the presence of GM ingredients.

10.6 Consumer expectations of GM labelling

Although the overwhelming majority of plant scientists recognise the use of GM as a legitimate research tool in support of current and future plant breeding, some farmers, consumers and environmentalists oppose the use of GM for a complex range of reasons.

The most common concerns expressed by those who oppose GM crops are: the potential for a negative impact on human health; possible contamination of other crops; the development of tolerance to pesticides with negative impact on the environment; and too much control exercised by big business.

All studies to date have found no greater risk from consumption of existing varieties of GM crops than from conventional crops. GM products intended for sale in the EU undergo very thorough safety assessments by the EFSA. In addition, GM technology has the potential to produce crops with positive health benefits, with existing examples including crops with enhanced nutritional content.

Genetic crossing of GM crops with non-GM crops or wild plants is possible, but the chances of this happening vary between species and the different pollination processes. Although co-existence measures are in place, some mixing cannot be completely ruled out due to farming, transport and trading practices, hence the agreement for a threshold on labelling, as part of the existing European legislative framework, and the need for adequate measures on LLP.

The potential environmental impact of GM technology is assessed on a case-by-case basis and research is on-going to develop crops that repel or are more resistant to insects and diseases so the need for pesticides and other chemicals can be reduced.

The control of the seed trade by large companies is an issue that can affect GM and non-GM seeds alike and can therefore be considered separately from the use of GM technology. Many argue that increasing the amount of public funding for GM research could be of benefit to consumers and farmers.

Public opinion and consumer views and arguments have been documented over the years through a series of consumer surveys and qualitative research in the UK and in Europe.

10.8.1 Trends in UK consumer surveys

In 2003 the FSA published an extensive report titled "Consumer views on GM food" as part of a 3 year public dialogue to "help deepen public understanding of all the issues surrounding GM." The research showed that most consumers did not have entrenched views (either pro or anti) on GM food, but there was a suspicion of GM products and a lack of clarity about potential consumer benefits. Most of the concerns were about the potential impact of GM crops on the environment and on potential long-term health effects of eating GM food.

The FSA continued to carry out research on consumer attitudes to genetic modification through its surveys, and in November 2009 it published the findings of a qualitative research project by the National Centre for Social Research (NatCen). Unsurprisingly, the study showed a very strong correlation between consumer attitudes to GM food and to science in general, with supporters of GM food seeing science as a way to improved quality of life and opponents worrying about the speed of scientific progress.

10.8.2 Eurobarometer

The EC has been monitoring the evolution of public opinion in the Member States since 1973 to inform its policies. A Special Eurobarometer survey conducted in 2010 was the seventh, since 1991, to ask questions on biotechnologies. Many of the 25,000

European citizens surveyed indicated increased scepticism towards genetic modification compared to 2005. More probing questions provided a less negative picture, with the majority of participants saying they would buy a GM product if it was healthier or contained less pesticide residues. As shown in previous surveys, European citizens are confident that applications of biotechnology will improve their quality of life, particularly when biotech is used for medicine and bio-based industrial applications, but remain unconvinced about GM applications to food production, expressing moral objections and concerns about potential risks.

10.7 Future trends

It is important that the mechanisms to assess the potential risks of different applications of biotechnologies to food production are robust, independent and fully trusted. The existing European regulatory framework has been recently reviewed and considered appropriate for protecting consumers and ensuring the efficiency of the market. The development of new selective breeding technologies will require future reviews of the existing regulatory framework and the implementation of existing regulation still remains a sticking point.

Despite an agreement on a solution on LLP for feed, the debate has not yet been had about the possible extension of the current legislation from feed to food. This situation will lead to unnecessary food waste if not properly and timely addressed by agreeing a threshold for events identified from crops that have received full approval in non-European countries.

Consumers and farmers in Europe will rightly continue to expect accurate information to be able to make informed choices about what they grow and eat. Discussions on the potential harmonisation of GM-free rules across Europe and on whether products from animals fed on GM feed should be labelled as such will continue.

The rising costs of raw materials and the challenges to food production from climate change coupled with increased food demand from a growing global population will also contribute to changing the context of future decisions on GM technologies in Europe. While other economies are already making the most of GM applications to food production, European Member States will need to find a balance between the practicalities of food security. Recent changes in retailers' policies in the UK with regard to the use of GM feed are a first indication of the position of Europe on a changing global market.

To conclude, a lot of effort should continue to be placed on providing trusted and unbiased information to consumers and on informing the debate on biotechnologies applications to food production with the most up to date and robust evidence. Only a good general and widespread understanding of the issues and an open forum for discussion can gradually allow the debate to happen without undue influence from over-polarised and entrenched positions.

10.8 Sources of further information and advice

Sources of information on GM are very broad and can be quite confusing, as a lot of the information is owned by stakeholders with a commercial or vested interest or by single issue groups with a very specific viewpoint.

Broadly, the information available is provided by government or public institutions, academia or independent organisations, commercial operators, trade bodies and NGOs.

Among the many sources of information from government or public institutions I would recommend WHO/FAO, the EC, the FSA in the UK, the Food and Drugs Administration (FDA) in the USA and the EFSA.

GMO Compass, SMC and Sense about Science provide information to demystify policies and explain the scientific evidence.

GM Freeze and the Soil Association are just two of the very active groups that provide information and run campaigns from an anti-GM perspective; the Crop Protection Association (CPA) and EuropaBio represent the views of commercial operators.

This is clearly a very superficial overview of just some of the available sources.

References

Regulation (EC) No 1829/2003
Regulation (EC) No 1830/2003
GM Timeline published on the Food Standards Agency (FSA) website
Briefing note published by the Science Media Centre (SMC) in July 2013
Science Media Centre Briefing Notes on GM Crops
http://www.sciencemediacentre.org/wp-content/uploads/2012/09/SMC-Briefing-Notes-GM-Crops.pdf
Europa
Foresight Future of Food and Farming report
The Future of Food and Farming (2011). Final Project Report. The **Government Office for Science**, London.
Directive 2001/18/EC
Commission Regulation EU No. 619/2011
In 2003 the FSA published an extensive report titled "Consumer views on GM food". In November 2009 it also published the findings of a qualitative research project by the National Centre for Social Research (NatCen)
Special Eurobarometer survey conducted in 2010
Eurobarometer previous survey in 2005

Websites

WHO – http://www.who.int/topics/food_genetically_modified/en/
FAO – http://www.fao.org/food/food-safety-quality/gm-foods-platform/en/

EC – http://ec.europa.eu/food/food/biotechnology/index_en.htm
EFSA – http://www.efsa.europa.eu/en/topics/topic/gmo.htm
FSA – http://www.food.gov.uk/policy-advice/gm/
FDA – http://www.fda.gov/Food/FoodScienceResearch/Biotechnology/
GMO Compass – http://www.gmo-compass.org/
Science Media Centre
Sense about Science
GMFreeze
Soil Association
Crop Protection Association
European Association for Bioindustries – http://www.europabio.org/
Grocery Manufacturers Association (GMA) – www.FactsAboutGMOs.org

Smart labelling of foods and beverages

11

G.A. Skinner
Insignia Technologies Ltd, Newhouse, UK

11.1 Introduction

The principle function of traditional food packaging involves the reduction of food deterioration (thereby extending its shelf life), maintenance of quality and safety, and the reduction/elimination of physical damage to food. Packaging serves to protect the contents from factors such as heat, light, moisture, pressure, oxygen, enzymes, micro-organisms, odours, insects, dust and dirt. The traditional secondary function of food packaging lies within the marketing of the product. Recent decades have increasingly demonstrated the importance of packaging as an inherent part of the marketing mix. Although a secondary function, this aspect of food packaging has a direct impact on the sales of the product. Interestingly, all traditional packaging endeavours to minimise the contact between food and packaging, maintaining the functions as described above.

However, the significant problem of food waste and increased consumer demands for mildly preserved, fresh tasting, convenient foods, along with changing retailing practices (i.e. globalisation of food distribution), more sophisticated packaging solutions have begun to emerge. There was an estimated 7.2 million tonnes of household food waste in the UK in 2010 (Quested and Parry, 2011). By allowing the consumer to make more informed decisions about the quality of the packaged food, it is hoped that less food will be thrown away unnecessarily. As the demand for high quality packaged food has continued, scientific research has made significant advances in the area of packaging technology. Thus, the interaction between food and packaging has begun to change, often with the aim of providing the consumer with real time information about the quality of the food contained within. This is the basis of intelligent packaging.

It is evident that there is some confusion surrounding the definition of intelligent packaging and it is often used interchangeably with similar, yet different, terms such as smart and active packaging. Yam *et al.* (2005) define intelligent packaging as "a packaging system that is capable of carrying out intelligent functions (such as detecting, sensing, recording, tracing, communicating and applying scientific logic) to facilitate decision making to extend shelf life, enhance safety, improve quality, provide information and warn about possible problems." They highlight that the primary aim of an intelligent packaging system is to *communicate* a problem, whereas

Advances in Food and Beverage Labelling. http://dx.doi.org/10.1533/9781782420934.3.191

an active packaging system is primarily concerned with taking *action* (e.g. release of antimicrobials) to protect the food product. Robertson (2006) defines active packaging as "packaging which subsidiary constituents have been deliberately included in or on either the package material or package headspace to enhance the performance of the package system." Examples of active packaging include the slow release of preservatives, the absorption of moisture or a component which maintains a desired temperature within the package. Although the term smart packaging is most commonly used as a broad definition which encompasses both intelligent and active packaging (Kerry and Butler, 2008), it is commonly used in place of intelligent packaging. Although the potential advantages of smart packaging are numerous, there have been concerns regarding the use of smart packaging. In an era where consumers are more aware of the environmental repercussions from using excessive non-biodegradable plastics, concerns have been raised regarding the increased quantity of packaging which might be required for smart packaging to be widely used (Butler, 2001). Additionally, chemical migration from the indicators into the foodstuffs has also been raised as a valid concern. Until 2004 there was no formal legislation on the correct usage and production of smart packaging in Europe. Regulation 1935/2004/EC and, more specifically, the most recent regulation 450/2009/EC, set out to define parameters for their development, safety and marketing. These articles stress the importance of migration testing on all products, ensuring that no chemical migration occurs from the intelligent/active packaging to the foodstuff. They also stipulate that such packaging must not mislead the consumer.

One of the most convenient ways of incorporating intelligent packaging technologies into food and beverage packaging is via smart labels. The functionality of smart labels is typically in the form of inks which are chemically formulated so as to interact with the surrounding environment. There are also examples of labels which rely on mechanical activation (e.g. Timestrip®, http://www.timestrip.com/), as well as the rapidly expanding electronic sector of smart labels. This short review will cover the main advances in smart labelling over recent years, covering the various approaches, methods and sensing technologies which have been used to develop a wide range of smart labels able to detect changes in temperature, freshness, oxygen concentration and carbon dioxide concentration. A brief review of labels using electronic technology will also be presented. It is obvious that this is an area of research which is rapidly expanding, and we can expect to see a rise in commercialised sophisticated smart labels appearing on various food and beverages in coming years.

11.2 Labelling to detect changes in temperature

Thermochromism is a property of substances which change colour due to a change in temperature. Uses of thermochromic compounds have been widespread throughout a range of industries including clothing, electronics, medical devices, toys, jewellery and novelty gifts; however, it has also found uses in smart labelling of food and beverages. The control of temperature in food production and transportation is extremely important to control the safety and freshness of the food product; hence, smart labels

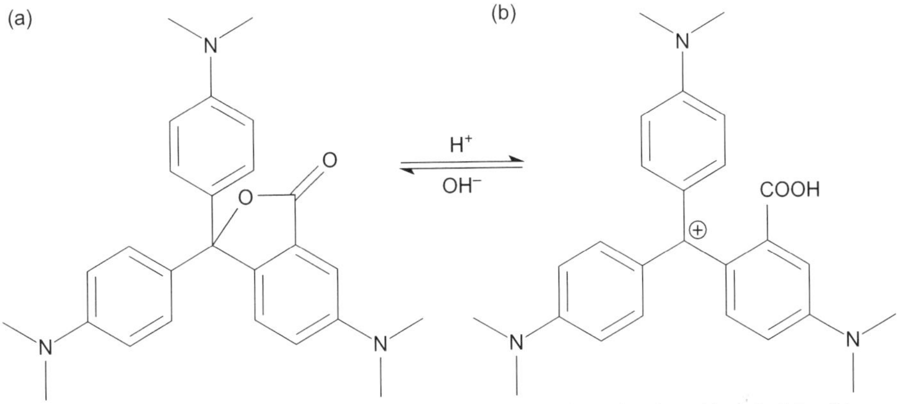

Figure 11.1 Chemical structure of crystal violet lactone in its colourless (a) and violet (b) forms.

which provide real time and historic information of the temperature storage are of great interest.

As well as sophisticated temperature monitoring, a number of temperature-sensitive labels have been applied to products as marketing tools. The labels seek to make the consumer more engaged with the brand than traditional labels. One of the most well-known examples of this is Coors Light Beers, where the mountains depicted in the background of the label changes from white to blue when the optimum drinking temperature is reached (see Plate VI in the colour section between pages 118 and 119).

Thermochromic inks are most commonly based on leuco dyes, which have the ability to change molecular structure in a way which changes their colour. One of the most well-known leuco dyes is crystal violet lactone, and its chemical structure is shown in Figure 11.1.

Upon a decrease in pH, the colourless lactone form (see Figure 11.1a) rearranges to its tertiary benzylic cation form (see Figure 11.1b),), gaining a strong colour (in this case violet). However, the temperature dependence of this colour change is due to the presence of a weak acid and a high molecular weight solvent (White and LeBlanc, 1999). The colour change takes place at a particular temperature (melting point of the chosen solvent). Interestingly, the colourless form of the dye is favoured at higher temperatures and the coloured form at cold temperatures, which suggests that the weak acid and the dye interact in the solid state, but are separated in the liquid state. However, the exact mechanism of how these interactions occur is not well known.

LCR Hallcrest (http://www.lcrhallcrest.com/) have developed a number of thermochromic food labels for a range of applications. Examples include labels for pizza boxes which turn from grey to pink, or black to red, above 45°C, thereby assuring the consumer that the pizza is still hot, labels for microwavable products which show that a required temperature has been reached, and labels for microwavable packs which change colour to warn consumers of very hot food (e.g. Hungry Jacks Maple Syrup).

They also report of a Ben and Jerry's cream tub marketed as "Mood Magic" which contained a thermochromic label allowing the consumer to interact with the packaging. The colour label would change colour with the person's body heat.

The main emphasis of these previous thermochromic labels has been to enhance the brand of the product rather than to provide confidence in the food quality. Time–temperature indicators (TTI) tend to have a more serious focus on providing the consumer with confidence that the product has been stored adequately prior to opening. TTIs are simple, cost-effective and user friendly devices to monitor, record and cumulatively indicate the overall influence of temperature history on the food product quality from the point of manufacture up to the consumer (Taoukis and Labuza, 1998). Of all smart packaging to date, TTIs have been the most successful on a commercial scale, with a number of supermarkets using them on their products.

Fresh-Check® TTIs are produced by TempTime Corporation, USA, and are specifically formulated to match the shelf life of the food products to which they are affixed (Fresh-Check®, http://www.fresh-check.com). The label consists of an oval shape containing a central circle which irreversibly darkens when the product is exposed to high temperatures. When the centre of the circle matches the colour of the oval the advice is to "use soon" and when the circle is darker than the colour of the oval, the product should not be consumed. Plate VII illustrates these colour changes.

The underlying chemistry of the colour change is a polymerisation reaction of a diacetylenic monomer (colourless), forming a dark coloured polymer (Smolander *et al.*, 1999). Exposure to heat drives the chemical reaction, hence its particular use in TTIs. Fresh-Check® labels have been used on products in Trader Joe's in the USA and Monoprix in France (Smolander *et al.*, 1999). Since Fresh-Check® TTIs are temperature-only activated (i.e. not mechanically activated), they obviously require deep-freeze storage prior to use. This feature is a major disadvantage to these indicators being widely commercialised, due to the extra cost involved in implementing these special storage conditions.

OnVu™ is a commercialised TTI currently being promoted by BASF and FreshPoint (OnVu™, http://www.onvu.de/). The OnVu™ label involves a colour change which the consumer compares to printed reference colours, as shown in Plate VIII. The sensing component of the label involves a colourless ink which turns dark blue when irradiated with UV light. The reverse reaction (dark blue to light blue to colourless) is driven by temperature and hence can be used to monitor the temperature history of the product. It is claimed that the label can be customised to suit various products and applications. Geissler's Supermarkets is reported to be the first US grocery retailer to adopt OnVu™ on meat product labels (Progressive Grocer, http://www.progressivegrocer.com/top-stories/headlines/meat-seafood/id34759/geisslers-adopts-onvu-meat-labeling/) and, if successful, it is hoped that it will be used on a wider range of products.

Cryolog have focused their efforts on creating a number of TTIs using predictive microbiology (http://www.traceo.com/en/compagnies/cryolog.html). A recent TTI by Cryolog has been approved for use by the European Food Safety Authority (EFSA) Panel on Food Contact Materials (EFSA, 2013). The indicator consists of the nonpathogenic micro-organism *Carnobacterium maltaromaticum*, within a nutritive gel medium, together with the pH-indicator, acid fuchsin. Depending on the time–temperature profile experienced, the *C. maltaromaticum* grows using the carbon source of the medium and as a result releases biogenic amines. The amines react with the pH-indicator causing an observable colour change from green to red. The indicator is contained within a multilayer polymer sachet which is adhered onto the outside of the packaging and is suitable for use on all types of chilled food stored at temperatures between 2°C and 12°C. This example demonstrates how microbiology and chemistry can be combined in order to create novel smart labels.

Although TTIs are excellent at providing information about temperature abuse of the *packaged* food products, they cannot accurately provide information about the freshness of the food, since they cannot account for any temperature abuse prior to packaging (e.g. improper storage at processing plant). In order to access the quality of the food, it is necessary to detect volatiles which correlate to the freshness of the food. This is the aim of freshness indicators.

11.3 Labelling to monitor freshness

As food degrades volatile gases are released due to the breakdown of proteins, fats and carbohydrates by bacteria, yeasts and fungi. The detection of these gases can provide an indication of the freshness levels of packaged food. Analytes which can be detected include carbon dioxide, ethanol and organic acids; however, the most commonly targeted analytes are volatile nitrogen compounds (e.g. ammonia, dimethylamine and trimethylamine) and biogenic amines (e.g. cadaverine and putrescine). Since amines can be easily detected via change in pH, the use of pH-sensitive dyes has been widely investigated in this area of smart labels. Equation [11.1] describes the interaction between the dye (D) and volatile amine which results in the observed colour change.

$$\underset{\text{(Colour A)}}{\text{DH} + \text{amine}} \rightleftharpoons \underset{\text{(Colour B)}}{\text{D}^- + \text{amineH}^+} \qquad [11.1]$$

where amineH$^+$ = protonated volatile amine.

Volatile amines are most prevalent in fish degradation and there have been a number of examples of smart labels for seafood applications. Pacquit *et al.* (2007) developed a smart label comprising a pH-sensitive dye entrapped in a polymer matrix, which colorimetrically indicates the spoilage of fish through the detection of amines. The results of this development work illustrate that the colour change of the indicator correlates well to the bacterial growth patterns in cod and whiting fish samples. Although not commercially developed, this work illustrates the potential for such indicators and the information they would be able to provide. Similarly, the Fraunhofer Institute in Germany have developed a colour-changing film which responds to biogenic amines released from fish products in order to track the freshness (Fraunhofer, 2011) (see

Plate IX). Although this work shows significant potential, there is no evidence to date of any commercialisation of this indicator.

The ripening process of fruit can be monitored by the detection of alkenes, alcohols, aldehydes and ketones (Bartley and Schwede, 1989). There have been examples of ripeness indicators which detect ethylene (Lang and Hübert, 2012); however, the indicators are often too costly, due to the need for expensive transition metal compounds. One commercial example of an intelligent ripeness label is RipeSense™, developed at the Horticulture and Food Research Institute of New Zealand. As observed from the picture in Plate X the indicator label starts off red, then graduates to orange and eventually yellow indicating "crisp," "firm" and "juicy," respectively (RipeSense™, http://www.ripesense.com). This smart label removes the need for consumers to squeeze and prod food, by providing real time information about the ripeness of the product. Perhaps one disadvantage of such smart labels is the requirement for additional packaging in order to trap the relevant volatiles. The RipeSense™ smart label has been sold in the UK, USA and Canada for use with pears, as well as in Germany for avocados. Research on new technologies is continuing in order to broaden their range of fruits.

A smart label that detects the ripeness of fermented vegetable foods, such as *kimchi*, was developed by Hong and Park (1999, 2000). Since commercially packed *kimchi* undergoes continual natural fermentation, the ability to continually monitor the ripeness of this product in a non-destructive manner, after packaging, is of high value. The authors report that titratable acidity (TA) and pH are key markers of the level of fermentation; hence, these ripeness indicators are based on pH-sensitive dyes, such as bromocresol purple, along with a CO_2 absorbent such as calcium hydroxide and polypropylene as the carrier resin. The resulting smart label provides a gradual colour change as the fermentation progresses, providing up-to-date information about the ripeness of the product.

Although CO_2 indicators are more commonly used as leak detectors, which are discussed in detail later in this review, examples of CO_2 indicators as freshness indicators are also to be found in literature. Nopwinyuwong *et al.* (2010) report an indicator for monitoring intermediate-moisture dessert spoilage. The indicator changes colour as the levels of CO_2 increase due to the gradual breakdown of the food by microbial growth. The indicator comprises a mixture of pH-indicating dyes resulting in a label which changes from green to red upon exposure to CO_2 in the region of 0.1–2.5% CO_2.

The detection of hydrogen sulphide (H_2S) has also been targeted in order to develop spoilage indicators. Ahvenainen *et al.* (1997) developed a freshness indicator for the quality control of modified atmosphere packed poultry meat. The indicator is based on the colour change of myoglobin by H_2S, which is released during the ageing of packaged poultry during storage. It is reported that the colour change of the myoglobin-based indicators corresponds to the deterioration of the quality of the poultry.

There are a number of approaches being investigated in order to develop smart labels which have the ability to accurately measure the freshness of packaged food. However, there are also a number of challenges (e.g. applicability across species, cost and customer acceptance) which still need to be addressed in order to see increased commercialisation of these potentially revolutionising technologies.

11.4 Labelling to detect changes in oxygen concentration

Aerobic microorganisms, which are the main cause of food spoilage, require oxygen to grow, multiply and degrade the food. Oxygen also allows many enzyme-catalysed reactions to take place e.g. the browning of food and vegetables, or the degradation of flavours. Fats in foods are susceptible to oxidation or rancidity, resulting in off-flavours and occasionally off-colours. Storing foods at refrigerated temperatures reduces these processes, but only increases their shelf-lives minimally. In order to overcome these problems, it has become common practice to pack foodstuff in oxygen-free environments, such as foods achieved with modified atmosphere packaging (MAP). Even after flushing the food package with inert gases (e.g. nitrogen and carbon dioxide), the residual level of oxygen in a package headspace is often still significant (2%) (Mills, 2005), due to poor gas flushing and the ability of food to retain oxygen. The oxygen level in the package can also increase over time due to inefficient sealing and the permeation of oxygen through the package material from the surrounding air. Although these increased levels of oxygen are minimal, they can significantly decrease the quality of the packaged food. In order to overcome these problems, many manufacturers have added oxygen scavengers to food packaging, usually in the form of a sachet or a label. These sachets are used to maintain the level of oxygen in a food package at 0.1% or less. Oxygen scavengers work by undergoing a chemical reaction to remove any oxygen present in the headspace of the package. The scavengers can lower and sustain the oxygen content to approximately 0.01%. Scavengers based on iron oxide have been hugely successful, proven by the fact that many food manufactures still utilise similar scavengers today (Cooksey, 2010). In order for such scavengers to work effectively, there has to be a satisfactory flush of inert gas, as well as a durable seal to ensure no leakage. Secondly, the packaging material has to be well chosen, typically a flexible polymer with a high oxygen barrier property. However, if the food package integrity was compromised to the extent that a leak was formed, the oxygen scavenger would be considered useless, since there would be a constant influx of oxygen. Such leaks are often not visible to the human eye and only evident when moulds or discolouration occurs within the package, i.e. after food spoilage has taken place.

Following these concerns, it is evident that there is a need for a leak detection system which would detect an increasing concentration of oxygen. There have been various examples of oxygen indicators reported in literature. Typically, these fall into two categories: colorimetric and lumophoric. Colorimetric-based oxygen indicators depend on an observable colour change, commonly due to an oxygen-binding reaction, a redox reaction or a light-activated redox reaction. Lumophoric indicators rely on a changing intensity or lifetime of the indicator, which normally decreases with increasing partial pressure of oxygen. The main disadvantage of lumophoric

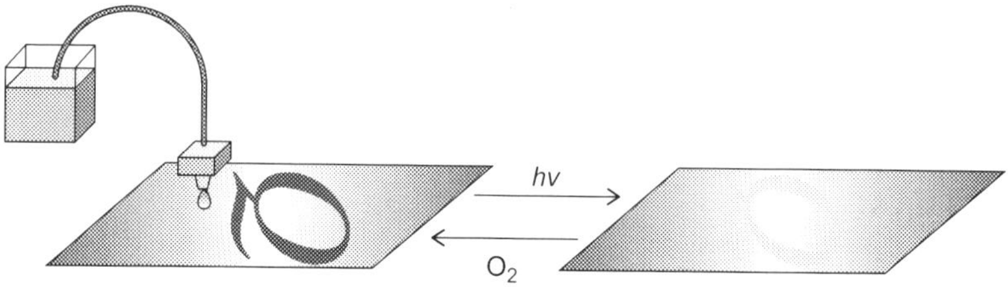

Figure 11.2 Illustration of printable oxygen-sensitive ink (Lawrie *et al.*, 2013).

indicators is the analytical equipment required to measure the level of oxygen present. Since colorimetric oxygen indicators rely only on an observable colour change, they are less expensive and can be easily interpreted by the consumer.

The colour change in colorimetric redox dye-based indicators is due to the different colours of the oxidised and reduced forms of the dye. Typically, the oxidised form of the dye is the most coloured species. The transformation from one form to the other can be carried out in two ways: using a chemical reducing agent (e.g. glucose in an alkaline medium (Yoshikawa *et al.*, 1979; Mattila-Sandholm *et al.*, 1998)) or via light-activation. Light-activation is typically carried out using a semi-conductor and a sacrificial electron donor. In brief, upon UV light irradiation of the nanocrystalline, finely dispersed, semi-conductor powder particles, electron-hole pairs are generated. These photogenerated holes are then able to irreversibly oxidise the sacrificial electron donor, allowing photogenerated electrons to accumulate on the semi-conductor particles. These electrons are then able to reduce the oxidised form of the dye to its reduced form (Mills, 2005). Both methods are effective at reducing a redox dye, causing a colour change. In both cases, oxygen is necessary to re-oxidise the reduced form of the dye, which makes these systems successful oxygen indicators. A number of redox dyes have been employed in a range of indicators, e.g. 2,6-dichloroindophenol (DCIP) (Eaton, 2002) and triphenyl tetrazolium chloride (TTC) (Ebraheem *et al.*, 2000); however, the most commonly used dye is methylene blue (MB). MB reversibly reverts from its colourless form (leuco-methylene blue (LMB)) back to its blue form. This distinct colour change provides the basis of a range of successful oxygen indicators.

A number of studies (Eaton, 2002; Lee *et al.*, 2005; Mills and Hazafy, 2009) have reported MB-based intelligent inks which are effective oxygen indicators. Lawrie *et al.* (2013) have recently developed an ink using colloidal TiO_2 opposed to dispersed TiO_2 nanoparticles, making it suitable for ink-jet printing. Once printed, the blue intelligent ink can be activated using UV light, which then reverts back to blue in the presence of oxygen, as illustrated in Figure 11.2.

Mills *et al.* (2012) reported "An O_2 smart plastic film for packaging," where oxygen sensitive pigments were dispersed in extruded low density polyethylene (LDPE). These indicators had similar sensing characteristics as previously reported indicators, but the flexible nature of the extruded polymer substrate allowed the indicator to be easily incorporated into existing food packaging. Both ink and pigmented film technologies have potential to be used as smart labels for the detection of oxygen;

however, they would most likely require further optimisation in order to create commercial products.

The Mitsubishi Gas Company developed the Ageless Eye™ oxygen indicator, intended to enhance the already established range of Ageless™ oxygen scavengers (MGC Inc., http://www.mgc.co.jp/eng/products/abc/ageless/eye.html). It relies on a reducing agent (glucose in alkaline conditions) to convert MB to the colourless form. In addition to MB, a non-redox dye, Acid Red 52, is usually added to provide a pink background colour (Yoshikawa *et al.*, 1979). The indicator components are pressed together to form a pellet which is then inserted into an oxygen-permeable sachet. Although this indicator has many desirable qualities which have led to it being produced commercially, there are several aspects of its design which limit its use. The cost of the indicator is expensive relative to a standard item of food packaging, and the requirement for storage under anaerobic conditions.

Despite significant technical developments in the area of oxygen indicators, only a small number have been developed commercially. However, with further product development and with increasing interest in intelligent packaging from the food industry, it is hoped that a number of these indicators will be developed into commercialised products in the near future.

11.5 Labelling to detect changes in carbon dioxide concentration

The detection of carbon dioxide by smart labels is useful for a number of applications, since it is both a spoilage product and used in high concentrations to create a modified atmosphere within food packaging. The choice of carbon dioxide as a flush gas is based upon its plentiful supply, and its antimicrobial action, i.e. it is an *active* packaging gas. The description "*active*" indicates that carbon dioxide inhibits microbe metabolism when present in high concentrations, even if oxygen is also present in the system (Brody *et al.*, 2001). Detection of levels of carbon dioxide in MAPed food is essential to indicate the package integrity, since any leak will cause increased microbial growth due to the increased oxygen content. Thus, accidental damage to the seal can cause the customer to buy spoiled food, without realising that the freshness has been compromised. By incorporating an intelligent colorimetric carbon dioxide indicating label into the package, the consumer would be able to know that the package integrity has been upheld and that the quality of food inside is still as estimated by the "use-by" date.

Carbon dioxide readily dissolves in water forming carbonic acid, which then dissociates to release hydrogen ions, see Equation [11.2]. Due to this acidic nature of carbon dioxide, the detection route for intelligent labels has primarily been via pH-sensitive indicators.

$$CO_2\,(aq) + H_2O \rightleftharpoons H_2CO_3 \rightleftharpoons H^+ + HCO_3^- \qquad [11.2]$$

For example, Balderson and Whitwood (1995) describe a reversible carbon dioxide indicator designed to be used as a leak indicator in MAPed food packages. Their patent describes a progressive colour change of five indicator strips as the concentration of carbon dioxide inside the package changes. The indicator was developed by Sealed Air Ltd., but does not appear to be on the market. There have been a number of academic research papers published describing colorimetric CO_2 indicators where use within food package labelling is suggested (Mills and Skinner, 2010; Mills *et al.*, 1992); however, few have been developed commercially.

Recently, Insignia Technologies Ltd. (http://www.insigniatechnologies.com) have commercialised intelligent pigment technology developed at the University of Strathclyde. These CO_2 sensitive colour-changing pigments can be dispersed and extruded into LDPE creating colour-changing plastics. Insignia Technologies Ltd, have subsequently used this technology to create an "embedded timer" which gradually changes colour after a pack of food has been opened (see Plate XI). The timing of the colour change is tailored to match the recommended use-by date once opened, thereby allowing the consumer to track how long food packages have been opened. Insignia are extending these smart labels for application in the cosmetic and medical markets.

Insignia have also used their pigmented film technology to develop a label for use in bulk soft fruit transportation. It is crucial that carbon dioxide levels throughout the transportation are maintained at 15% CO_2 to ensure optimum quality. However, until now it was difficult to accurately know if such levels were being maintained, or if a leak was present, without the use of expensive analytical equipment. Leakage in the packaging causes the carbon dioxide level to significantly decrease, resulting in a poor quality of fruit by the time the produce reaches the supermarkets. Insignia's label is applied to the inside of the fruit pallet by packers before the fruit is filled with the modified atmosphere gas and sealed. As illustrated in Plate XII, the colour-changing "tick" is yellow when the CO_2 concentration is at the recommended level and changes to purple if the CO_2 concentration drops below the recommended level. Using this simple smart label, any leak present is highlighted before the fruit gets delivered to the customer.

11.6 The use of electronic technology to develop smart labelling

The development of Radio Frequency Identification (RFID) tags has opened up the potential for food products to be tracked throughout the supply chain, providing substantially more up-to-date information about the products than barcodes. RFID tags have found use in many applications within the food and beverage industry, for food monitoring and traceability, improving supply chain efficiency and enhancing food safety, etc.

RFID tags contain digital data encoded in the tag and are captured by a reader via radio waves (Yam *et al.*, 2005). Typically, the tags consist of an integrated circuit, an antenna and a battery, if the tag is active. Passive tags do not require a battery since they are powered by the energy supplied by the reader. Once captured, these data are then passed to a host computer, where they can be analysed and relevant decision making carried out. RFID is similar to barcoding in that data from a tag or label are captured by a device that stores the data in a database. RFID, however, has several advantages. Most notably, RFID tag data can be read remotely outside the line-of-sight, whereas barcodes must be aligned with an optical scanner. The reading range of the RFID depends on many factors, including the frequency of operation and the power of the reader. RFID tags are capable of storing a significantly larger quantity of data compared to barcodes. However, the major weakness of RFID technology is the susceptibility for some types of packaging and food products to interfere with the RFID signal (Clarke *et al.*, 2006; Kumar *et al.*, 2009). Metals can have a significant shielding effect, which makes the RFID signal difficult to detect. However, it has recently been discovered that low frequency RFID (125–135 kHz) has a reduced loss of radio signal by metal in the high frequency magnetic field of a reader compared to higher frequency signals (Brody *et al.*, 2008) in the presence of metal packaging. The presence of water is another major problem, which causes detrimental effects on readings since water molecules can absorb energy, causing the tag to detune. A number of suggestions to overcome both these problems are reported throughout literature (e.g. Clarke *et al.*, 2006). However, in order for RFID tags to be widely used throughout the food packaging industry, the tags must be sufficiently versatile to cope with a range of environments.

High cost and poor recyclability are other major hurdles RFID technology will face in order to develop widespread use within the food industry. However, with increasing demand for traceability within the supply chain, the use of RFID is likely to increase. In 2003, Walmart issued a mandate requiring its top 100 suppliers to use RFID on all pallets entering their distribution centres by 2005 (Roberti, 2004). Despite not achieving this target, it is evident that Walmart contributed significantly to the increasing use of RFID throughout the retail sector and are still dedicated to the wide use of RFID on their products.

In recent years it has become increasingly common to find quick-response (QR) codes on food and beverage packaging. QR codes are essentially matrix barcodes that can be scanned by most smartphones, which leads the consumer to a website relevant to the brand. QR codes can provide the consumer with an additional layer of information and increases the interaction between the consumer and the product. While the uptake of this technology is considerable across a wide range of brands, views regarding the future of QR coding are mixed (Gyeke, 2013). Some experts predict that QR codes will disappear as quickly as they appeared, while others predict that the technology will continue to become more sophisticated and has a promising future.

Another key focus of electronic smart labels is to provide an "entertainment experience" for the consumer, thereby enhancing the brand of the product and the consumer's connection with one particular brand. An example of this includes Heineken's Ignite beer bottle, which lights up when in contact with other Heineken bottles and can actively respond to the surrounding environment (e.g. sparks when the consumer drinks, and "sleeps" when stationary). Additionally, the bottle can be remotely activated, allowing all bottles to become an active light source synchronised to a music beat (Heineken, 2013).

Although electronic smart labels are among the most complex and expensive, they provide significantly more information than other smart labels, making them extremely valuable to progress the provision of information via smart packaging.

11.7 Conclusions and future trends

A wide variety of smart labelling technologies exist; however, the majority of products are yet to make significant impact on the packaging industry. Ahvenainen (2003) suggests that intelligent packaging have a "bright future" due to the increasing significance of food freshness and safety, increasing consumer demands, as well as increased globalisation and expansion of marketing, leading to longer logistic chains and hence more demands on food traceability. Recent studies (Anon, 2012) have suggested that the world demand for food safety products is anticipated to rise above 8% per year, reaching $18bn USD in 2016. The study suggests that this demand is driven by more stringent food safety regulations, as well as by a number of recent high profile international food related illness outbreaks. A similar study (Dewan, 2013) notes that active packaging is worth $8.8bn USD, and is set to grow by 5.2% annually to $11.9bn USD. Similarly, intelligent packaging is suggested to grow from $3.8bn USD in 2011 to $5.3bn USD in 2017.

Interestingly, the report also notes that it is vital that manufacturers communicate thoroughly with consumers when introducing new intelligent and active packaging solutions to ensure continued acceptance of these products. As well as effective communication, a number of other factors must be addressed in order for smart packaging to grow at the predicted rates. Ease of large-scale manufacture of the smart labels is critical to product success and must be addressed at the initial stages of product development. The additional cost of using smart labels is a major factor which often deters uptake; therefore any new smart label for fast moving consumer goods must add minimal cost to the overall packaging. As with any new technology, there is always a degree of risk by adding a new feature on an established product. However, this should be minimised by ensuring all products perform to a high level and that adequate customer education about the functionality is always carried out. The application of the smart labels to the packaging ideally must involve current label application methods. Alteration of the packaging process, or additional capital investment in order to purchase new applicators could potentially dissuade mass uptake of smart labels. Finally, the commercial benefits of the indicators must be clear throughout the supply chain.

Smart labelling has the potential to make a significant impact to the packaging industry and it is hoped, with time, that more of these revolutionising technologies will become commonplace on a wide variety of food and beverage packaging.

11.8 Sources of further information and advice

The Active & Intelligent Packaging Industry Association (AIPIA, http://www.aipia.info/) seeks to aid the commercialisation of active and intelligent packaging solutions within the food, pharmaceutical and cosmetic sectors. AIPIA provides up-to-date information on new developments within the area of smart labels.

Key books which address the area of smart packaging include:

Ahvenainen, R. (2003) *Novel Food Packaging Techniques*, Cambridge, Woodhead Publishing Limited.

Kerry, J. and Butler, P. (2008) *Smart Packaging Technologies for Fast Moving Consumer Goods*, Chichester, John Wiley & Sons Ltd.

References

Active & Intelligent Packaging Industry Association (AIPIA). Website information available from: http://www.aipia.info (Accessed 11 September 2013).

Ahvenainen, R. (2003), "Active and intelligent packaging," in Ahvenainen, R., *Novel Food Packaging Techniques*, Cambridge, Woodhead Publishing Limited, 5–21.

Ahvenainen, R., Pullinen, T., Hurme, E., Smolander, M. and Siika-Aho, M. (VTT Biotechnology and Food Research, Espoo, Finland) (1997), *Package for Decayable Foodstuffs*. PCT International patent application WO 9821120.

Anon (2012), World Food Safety Products to 2016, Ohio, Freedonia Group.

Balderson, S. N. and Whitwood, R. J. (Trigon Industries Limited) (1995), *Gas Indicator for a Package*. US Patent 5439648.

Bartley, J. P. and Schwede, A. M. (1989), "Production of volatile compounds in ripening kiwi fruit (Actinidia chinensis)," *J. Agric. Food Chem.*, **37**(4), 1023–1025. DOI: 10.1021/jf00088a046.

Brody, A. L., Bugusu, B., Han, J. H., Koelsch Sand, C. and McHugh, T. H. (2008), "Innovative food packaging solutions," *J. Food Sci.*, **73**(8), R107–R116. DOI: 10.1111/j.1750-3841.2008.00933.x.

Brody, A. L., Strupinsky, B. R. and Kline, L. R. (2001), *Active Packaging for Food Applications*, Lancaster, Technomic Publishing Co.

Butler, P. (2001), "Smart packaging – intelligent packaging for food, beverages, pharmaceuticals and household products," *Mater. World*, **9**(3), 11–13.

Clarke, R. H., Twede, D., Tazelaar, J. R. and Boyer, K. K. (2006), "Radio frequency identification (RFID) performance: the effect of tag orientation and package contents," *Packag. Technol. Sci.*, **19**(1), 45–54. DOI: 10.1002/pts.714.

Commission Directive 1935/2004/EC of 27 October 2004 on materials and articles intended to come into contact with food and repealing Directives 80/590/EEC and 89/109/EEC. L 338, 13.11.2004, 4–17.

Commission Directive 450/2009/EC of 29 May 2009 on active and intelligent materials and articles intended to come into contact with food. L 135, 30.05.2009, 3–11.

Cooksey, K. (2010), "Oxygen scavenging packaging systems" in *Encyclopaedia of Polymer Science and Technology*, John Wiley & Sons Ltd, West Sussex, England.

Cryolog. Website information available from: http://www.traceo.com/en/compagnies/cryolog.html (Accessed 23 August 2013).

Dewan, S. S. (2013), *Active, Controlled, and Intelligent Packaging for Foods and Beverages*: report no. FOD038C, BCC Research, USA.

Eaton, K. (2002), "A novel colorimetric oxygen sensor: dye redox chemistry in a thin polymer film," *Sens. Actuators B-Chem.*, **85**, 42–51. DOI: http://dx.doi.org/10.1016/S0925-4005(02)00048-5.

Ebraheem, S., Abdel-Fattah, A., Said, F. and Ali, Z. (2000), "Polymer-based triphenyl tetrazolium chloride films for ultraviolet radiation monitoring," *Radiat. Phys. Chem.*, **57**, 195–202. DOI: http://dx.doi.org/10.1016/S0969-806X(99)00352-7.

EFSA (2013), "Scientific Opinion: Scientific Opinion on the safety evaluation of a time-temperature indicator system, based on Carnobacterium maltaromaticum and acid fuchsin for use in food contact materials," *EFSA J.*, **11** (7), 3307. DOI: 10.2903/j.efsa.2013.3307.

Fraunhofer (2011), Research News: Rotten meat doesn't stand a chance, Fraunhofer Research Institution for Modular Solid State Technologies EMFT. Available from: http://www.fraunhofer.de/content/dam/zv/en/press-media/2011/pdfs/rn4_2011_APRIL.pdf (Accessed 13 August 2013).

Fresh-Check®. Website information available from: http://www.fresh-check.com (Accessed 11 September 2013).

Gyeke, L. (2013), The Big Question – Does the QR code have a future? Available from: http://www.packagingnews.co.uk/news/the-big-question-does-the-qr-code-have-a-future/ (Accessed 13 November 2013).

Heineken (2013), *Heineken Ignite Background.* Available from: http://files.tribalddb.nl/spark/BLOG POSTS/2013-05-03 Tribal DDB Heineken Ignite Talking points.pdf (Accessed 23 August 2013).

Hong, S.-I. and Park, W.-S. (1999), "Development of color indicators for *Kimchi* packaging," *J. Food Sci.*, **64**(2), 255–257. DOI: 10.1111/j.1365–2621.1999.tb15877.x.

Hong, S.-I. and Park, W.-S. (2000), "Use of color indicators as an active packaging system for evaluating *kimchi* fermentation," *J. Food Eng.*, **46**(1), 67–72. DOI: http://dx.doi.org/10.1016/S0308-8146(00)00141-2.

Insignia Technologies Ltd. Website information available from: http://www.insigniatechnologies.com/(Accessed 11 September 2013).

Kerry, J. and Butler, P. (2008), *Smart Packaging Technologies for Fast Moving Consumer Goods*, Chichester, John Wiley & Sons Ltd.

Kumar, P., Reinitz, H. W., Simunovic, J., Sandeep, K. P. and Franzon, P. D. (2009), "Overview of RFID Technology and Its Applications in the Food Industry," *J. Food Sci.*, **74** (8), R101–R106. DOI: 10.1111/j.1750-3841.2009.01323.x.

Lang, C. and Hübert, T. (2012), "A colour ripeness indicator for apples," *Food and Bioprocess. Techn.,* **5**(8), 3244–3249. DOI: 10.1007/s11947-011-0694-4.

Lawrie, K., Mills, A. and Hazafy, D. (2013), "Simple inkjet-printed, UV-activated oxygen indicator," *Sens. Actuator B-Chem.,* **176**, 1154–1115. DOI: 10.1016/j.snb.2012.10.125

LCR Hallcrest. Website information available from: http://www.lcrhallcrest.com (Accessed 12 August 2013).

Lee, S-K., Sheridan, M. and Mills, A. (2005), "Novel UV-activated colorimetric oxygen indicator," *Chem. Mater.*, **17**, 2744–2751. DOI: 10.1021/cm0403863.

Mattila-Sandholm, T., Ahvenainen, R., Hurme, E. and Jarvi-Kaariainen, I. T (VTT Technical Research Centre of Finland) (1998), *Oxygen-Sensitive Colour Indicator for Detecting Leaks in Gas-Protected Food Packages*, European patent application EP 0 666 977 B1.

Mills, A. (2005), "Oxygen indicators and intelligent inks for packaging food," *Chem. Soc. Rev.*, **34**, 1003–1011. DOI: 10.1039/b503997p.

Mills, A., Chang, Q. and McMurray, N. (1992), "Equilibrium studies on colorimetric plastic film sensors for carbon dioxide," *Anal. Chem.*, **64**(13), 1383–1389. DOI: 10.1021/ac00037a015.

Mills, A. and Hazafy, D. (2009), "Nanocrystalline SnO_2-based, uvb-activated, colourimetric oxygen indicator," *Sens. Actuators B-Chem.*, **136**, 344–349. DOI: http://dx.doi.org/10.1016/j.snb.2008.12.048.

Mills, A., Lawrie, K., Bardin, J., Apedaile, A., Skinner, G. A. and O'Rourke, C. (2012), "An O_2 smart plastic film for packaging," *Analyst*, **137**(1), 106–112. DOI: 10.1039/c1an15774d.

Mills, A. and Skinner, G. A. (2010), "Water-based colourimetric optical indicators for the detection of carbon dioxide," *Analyst*, **135**(8), 1912–1917. DOI: 10.1039/C000688B.

Mitsubishi Gas Company, AGELESS EYE Oxygen Indicator. Website information available from: http://www.mgc.co.jp/eng/products/abc/ageless/eye.html (Accessed 17 August 2013).

Nopwinyuwong, A., Trevanich, S. and Suppakul, P. (2010), "Development of a novel colorimetric indicator label for monitoring freshness of intermediate-moisture dessert spoilage," *Talanta*, **81**, 1126–1132. DOI: 10.1016/j.talanta.2010.02.008.

OnVu™. Website information available from: http://www.onvu.de/ (Accessed 14 August 2013).

Pacquit, A., Frisby, J., Diamond, D., Tong Lau, K., Farrell, A., Quilty, B. and Diamond, D. (2007), "Development of a smart packaging for the monitoring of fish spoilage," *Food Chem.*, **102**, 466–470. DOI: http://dx.doi.org/10.1016/j.foodchem.2006.05.052.

Progressive Grocer (2012), "Geissler's Adopts OnVu Meat Labeling." Available from: http://www.progressivegrocer.com/top-stories/headlines/meat-seafood/id34759/geisslers-adopts-onvu-meat-labeling/ (Accessed 19 August 2013).

Quested, T. and Parry, A. (2011), *New Estimates for Household Food and Drink Waste in the UK*, WRAP, UK.

RipeSense®. Website information available from: http://www.ripesense.com/ (Accessed 14 August 2013).

Roberti, M. (2004), Wal-Mart Begins RFID Rollout. Available from: http://www.rfidjournal.com/articles/view?926 (Accessed 14 August 2013).

Robertson, G. L. (2006), *Food Packaging – Principles and Practise*, New York, Woodhead Publishing Limited.

Smolander, M., Hurme, E. and Ahvenainen, R. (1999), "Smart packaging for increased product safety," Tech. rep., VTT Biotechnology and Food Research, Espoo, Finland. Available from: http://www.nordicfoodpack.org/16044 (Accessed 26 August 2013).

Taoukis, P.S. and Labuza, T. P. (1998), "Applicability of time-temperature indicators as shelf life monitors of food products," *J. Food. Sci.*, **54**(4), 783–788. DOI: 10.1111/j.1365–2621.1989.tb07882.x.

Timestrip®. Website information available from: http://www.timestrip.com/ (Accessed 23 August 2013).

White, M. A. and LeBlanc, M. (1999), "Thermochromism in Commercial Products," *J. Chem. Educ.*, **76**(9), 1201. DOI: 10.1021/ed076p1201.

Yam, K. L., Takhistov, P. T. and Miltz, J. (2005), "Intelligent packaging: concepts and applications," *J. Food Sci.*, **70**(1), R1–R10. DOI: 10.1111/j.1365–2621.2005.tb09052.x.

Yoshikawa, Y., Nawata, T., Goto, M. and Fujii, Y., (Mitsubishi Gas Chemical Co., Inc.) (1979), *Oxygen Indicator*. US Patent 4169811.

Labelling relating to natural ingredients and additives

12

S. Osborn
Leatherhead Food Research, Leatherhead, UK

12.1 Introduction

There is no doubt that the challenge presented when looking at moving to "clean label" formulae is the absence of a legal definition. This not only presents significant hurdles in terms of labelling, but also in the approach to reformulation in order to achieve it. The clean label proposition is generally considered to be seeking natural alternatives to food additives, and listing named ingredients on labels rather than E-numbers. The reality, however, is that this does not extend far enough, since "clean label" goes beyond E-numbers and additives and is increasingly seen to include ingredients themselves. This is leading to a crossover between what was previously segmented in terms of the type of fat, sugar and starch now being subjected to the sort of scrutiny that additives were previously and, as a result, in origin and processing becoming items of consumer focus.

Alongside the evolution of "clean label" has been the increase in the desire for "natural" ingredients, which has been a macro-trend over the last few years and continues to be of importance. It plays a significant part in the clean label discussion and, while it might be considered a separate derivative of clean label, it is now being widely considered to be identical.

The complexity and evolution of this topic, and the challenges it presents, are so great that they cannot all be explored here, but it is hoped that an understanding of the philosophies and approaches required can be provided.

12.2 Clean label definitions

"Clean label" is yet to be defined; there have been, however, various attempts to clarify the situation, which have tended to centre on the application of the term "natural." In the USA, the Food and Drug Administration (FDA) has indicated that "natural" may be used provided it is truthful and not misleading (i.e. the product does not contain any artificial flavour or flavourings, colouring ingredients, chemical preservatives or any other artificial or synthetic ingredients) and the product and its ingredients are not more than minimally processed. The Food Standards Agency (FSA) in the UK

Advances in Food and Beverage Labelling. http://dx.doi.org/10.1533/9781782420934.3.207

also provides similar guidance for the use of the terms "fresh," "pure," "natural," etc., in food labelling (Food Standards Agency, 2008). However, both these agencies tend to deal with the use of "natural" in labelling from a marketing perspective, and not the full "clean label" picture.

A full definition must reach much further. In 2010, Ingredion tried to achieve this by extending the scope to include label simplicity and minimal processing (Scott, 2010).

They suggested a three point definition:

- *Free from additives*: remove or replace food additives
- *Simple ingredient listing*: choose recognisable ingredients that do not sound chemical or artificial
- *Minimally processed*: process foods using traditional techniques that are understood by consumers and not perceived as being artificial.

This still does not, however, address the full sentiment, since it does not allow for the use of acceptable recognisable "artificial" additives, and it does not make allowance for the "in use" context and expectation of an ingredient to be present. Although it could be argued that this is addressed through the third point, in essence this does not address the consumers' *perception* sufficiently.

12.3 Clean label and the consumer

In 1964 a unified list for colours was assigned a number derived from the *Codex Alimentarius* International Numbering Scheme. This was expanded to include other categories during the following decade (Osborn, 2014). This was done in good faith to show consumers that additives had been tested and approved. This mechanism also provided structure and categorisation that formed part of the EU General Regulation (EC) 178/2002 (as amended) and the EU Labelling Directive 2000/13/EC (as amended), which dictate that food needs to be safe to eat and labelling must not be misleading. Variations in the use of the nomenclature can be found throughout the world to varying degrees.

However "E-numbers" have not been viewed in a positive light, and have been seen as an opportunity for the food and drink industry to hide the use of unnecessary components.

It is generally accepted that "clean label" involves the removal of chemical-sounding ingredients, artificial additives and E-numbers, and increasingly extends to the creation of healthier labels with the removal or reduction of salt, fat and sugar. This shift has also brought into usage the term "store cupboard" ingredients which, in its most simple form, can be considered as ingredients that the consumer recognises, while also understanding their function in the product. While this sentiment is ideal, the practicality of its implementation is somewhat different, and this creates the "clean label" paradox: consumers are setting the conditions for product profiles, but they do not really know what the rules are. However, the strong consumer message centres on the ability of a food or beverage product label to facilitate informed decisions and choices based on the products declared ingredients (Berry, 2011).

This ever-changing need provides the food and beverage industry with an opportunity to connect with the consumer on issues that are driving their purchasing habits. Food choice is emotive, and consumers are increasingly mistrustful of the unknown; thus the replacement with familiar ingredients is viewed positively amongst the purchasing public. An important element of this is that the customer is often not the consumer. In this scenario the customer assumes the responsibility of choice for the consumer. This proxy arrangement underpins much of the clean and natural choice, since the customer is most likely to make what is seen as a positive health choice for their charges. This is particularly the case in mothers buying for children.

12.4 Clean label/Natural/Free from

As we have indicated, clean label is driven by a consumer need to understand a food label by recognising the ingredients. This can, however, conflict with "naturalness," since certain ingredients could be considered "natural" but not clean label because of a lack of familiarity. A reasonable example of this is xanthan gum (E415) – it is considered by many to be "natural" since it is derived from the fermentation of sugars such as sucrose; however, it is not an ingredient typically found "in store cupboards" and would not necessarily be recognised by a consumer, and thus it might not be considered "clean label." This was reported in consumer studies by Leatherhead Food Research (Beeren, 10), who suggested that xanthan gum (E415) is viewed with scepticism, not because of its fermented origins, but because of its name; similarly, guar gum (E412), a natural polysaccharide, is also treated with some concern.

One of the significant shifts in clean labelling declarations has been seen with the use of "free from" statements on-pack. This has been used to circumvent the complexities of what is/is not natural, since extolling the absence of artificial ingredients is less contentious, yet conveys the same sentiment. The marketing message that a product "free from artificial colours/flavours," etc., is widespread, and delivers a clear communication in order to meet the consumer needs. Such labelling with colours and flavours is possible since the widely publicised issue with artificial colours in the "Southampton 6 Report" (McCann *et al.*, 2007) has resulted in a clear demarcation between artificial and natural colours, and Regulation (EC) No. 1334/2008 on flavourings and certain food ingredients with flavouring properties for use in and on foods. This legislation not only includes flavourings and their source, but also food ingredients with flavouring properties. A "flavouring," in this Regulation, aligns with that of *Codex Alimentarius*: "Products not intended to be consumed as such, which are added to food in order to impart or modify odour and/or taste." In this context, "natural" can only be used when the flavouring component comprises only natural flavouring substances or preparation. This requires at least 95% to be derived from the named source, and the remainder is not to taste the named source but only to provide standardisation or enhancement. A declaration that "natural X flavouring with other natural flavouring" is permitted when the flavouring component is less than 95% (Osborn, 2012).

This inclusion of "free from" is further complicated since its evolution, in an additive context, has occurred alongside the introduction of allergen labelling in Europe (Directives 2003/89/EC and 2006/142/EC (to be superseded in December 2014 with the implementation of the Food Information for Consumers Regulation (EU) No. 1169/2011)), and the increase in products targeted at sensitive populations. This results in some instances of clean label reformulation extending to include gluten free, lactose free, etc. However, since such sensitive populations are in general in the minority, "allergen free" products should not be considered within the scope of clean label strategies.

12.5 Clean label challenges

Clean label is holistically ideal, in that it delivers the opportunity for product developers to respond to consumer demand, enables them to reconnect with food and the ingredients, provides a perceived healthier proposition and, most importantly, helps to regain the trust between "industrial" entities and consumers. Since there is deep cynicism amongst some consumers regarding large scale food and beverage processing, this mechanism to regain trust is essential. However, the move towards cleaner labels can provide a series of challenges, and stability and shelf life must be considered the primary issues.

Colours have had particular technical difficulty in making the transition to natural alternatives. Ingredients such as "beetroot red" (E162), which deliver clean label credentials, are less stable, pH sensitive and may not be as widely applicable as the artificial alternatives. This requires a close relationship and dialogue with colour suppliers to ensure that the "natural" alternative selected is actually suitable for your product matrix. The stability of the colour, and in particular its photosensitivity, is of paramount importance; after all, it is the fundamental feature of the product's visual impact, and only through careful shelf life studies can the appeal of the product throughout its life be assured.

This forms one of the great paradoxes of the clean label movement, since modern food and beverage production has built much success on stable and homogenous products, and it will take some time for consumers to readjust to acceptable variation within products. This is also leading to exploration into technologies such as encapsulation and nanotechnologies that may enhance colour stability (Teoh, 2012). However, this precipitates the need for a clear philosophy to ensure that any innovations are aligned to the consumers' expectations of what should be in their product.

A further challenge arises with the use of preservatives: additives such as sorbates and benzoates are common, but have also been the focus of studies that suggest they play a role in attention deficit hyperactivity disorder (ADHD) and asthma. This has fuelled an appetite for their removal; however, in many cases at the time of writing, effective alternatives are yet to be found. The use of additives in the preservation of food and beverages is essential to ensure food safety and product integrity, and their wholesale removal has significant impact on shelf life. It is therefore the focus of much research, and recent additions to the EU positive list include a clean label antioxidant, "extract of rosemary" (E 392). Novel processes, such as high pressure processing, are increasingly being explored as less aggressive, non-thermal inactivation processes. However, investment in such technology can be large, and it may also rest uneasily in the eyes of the consumers, who may not consider such novel processing to satisfy the "minimal processing" test.

Once a clear direction of travel in removing unnecessary ingredients and label simplification has been established, the overarching driver in clean label relates to the use of recognised ingredients. It is therefore not enough simply to claim a clean label origin. For example, fermentation may be considered a natural process and can derive ingredients with preservative properties such as propionic acid (Sahi, 2012), but this provenance is unfortunately not conveyed on the label and is indistinguishable from any other source, which may result in an exclusion from the clean label toolkit, albeit unjustly.

With these challenges in mind, it is essential to develop a clear philosophy with regard to approaching clean label reformulation. This philosophy must encompass a broad range of factors, including safety, stability, acceptability, functionality, ease of manufacture and cost in use (Morley *et al.*, 2013). Implicit within this is the importance of a degree of flexibility to ensure that these are not compromised and that the consumer is engaged with the process so that any compromises are accepted, as reformulation is often a balancing act and the need for transparency and honesty is paramount within a clean label strategy (Osborn, 2014). Diaz (2013) advocates a systematic science-based approach to clean label reformulation, stressing the importance of understanding the food matrix and ingredient interactions. This enables the elimination, modification or replacement of ingredients as required from an informed standpoint. This approach is essential in overcoming the challenges and achieving consumer and regulatory acceptability.

12.6 Clean label ingredients

As has been outlined, the need to adopt a clean label approach is driven by the consumers and is a well-established trend. Wyers and Tremellen (2013) allude to such values as part of Innovas Market Insights Top Trends for 2014, referring to trust, simplicity and alternatives in three of the ten themes, all key components of achieving a clean label.

The landscape of ingredients is complex, and it is worth considering each of the ingredient categories and some of the clean label options that exist within them.

Within the EU, the use of additives in certain foods is only permitted to specific quantitative limits. Additives are categorised by the FSA (2014) as follows:

- Colours
- Preservatives
- Antioxidants
- Sweeteners

- Emulsifiers, Stabilisers, Thickeners and Gelling Agents
- Others
  - Acid, acidity regulators, anti-caking agents, anti-foaming agents, bulking agents, carriers and carrier solvents, emulsifying salts, firming agents, flavour enhancers, flour treatment agents, foaming agents, glazing agents, humectants, modified starches, packaging gases, propellants, raising agents and sequestrants.

Further to these additives, the macro-ingredients (fat, carbohydrate, protein and fibre) are also subject to clean label scrutiny.

12.6.1 Colours

Colours are an additive category that have had a particularly high profile over recent years, particularly in relation to "clean label" and "natural." The "Southampton 6 Report," based on research from McCann *et al.* (2007), highlighted potential issues relating to hyperactivity in children and suggested correlation to the use of six artificial colours: sunset yellow (E 110), quinoline yellow (E 104), carmoisine (E 122), allura red (E 129), tartrazine (E 102) and ponceau 4R (E 124). These have been replaced with a range of alternatives which are considered to deliver "clean label" credentials. These alternatives, however, do come with stability issues when exposed to heat, light or changing pH. Some of these are broadly summarised below (Table 12.1).

12.6.2 Preservatives

The scope for natural preservatives is broad, with the expectation for them to have a wide antimicrobial action, be applied at low concentration, not contribute sensory impact to the product and be generally stable, and it is widely recognised that no single solution can achieve all this (Delves-Broughton, 2012). Preservation is a fundamental of food production to ensure the stability and safety of a food or beverage product, and potentially provides the greatest challenge to providing a recognisable ingredient-led label. Typical preservatives have been benzoates and sorbates, and while a range of fermentation-derived alternatives exist, such as nisin (antimicrobial) and natamycin (antifungal), there is a need to produce a substantial level of consumer education for these to be understood and accepted.

The preservation paradigm demonstrates the need for inclusive dialogue with consumers in order to gain a greater understanding of their expectations and to provide information and education at the same time. This may then lead to processes that can provide essential preservation functions, such as high pressure processing, pulsed electric field technology, or irradiation, even without the need for additive use. Irradiation (whereby food is exposed to ionising radiation) is a process which could result in a simplified ingredients list. However, negative associations of radiation, the requirement in some territories such as the USA for products to carry relevant declarations, and major retailer policies prohibiting its use have impacted on its widespread uptake. This is despite detailed studies by World Health Organization (WHO), United Nations Food and Agriculture Organization (FAO), European Community Scientific Committee for Food, and the United States FDA, and a 2011 statement by

Table 12.1 **Stability of natural colours**

Colour	Colour shade	pH stability	Heat stability	Light stability	Oxygen stability
Anthocyanin	Red to blue (depends on pH)	>pH 3.8 colour changes	Good	Fair to good	Sensitive
Annatto	Yellow to orange	Precipitates at < pH 4; degrades in acids	Fair	Fair to good	Sensitive
Beta-carotene	Yellow-orange to orange-red	Good at pH 2–8	Good	Fair to good	Sensitive
Caramel	Brown	Good (stable at pH 3–10)	Good	Good	Good
Carmine	Orange to red	Good	Good	Good	Good
Chlorophyll	Olive green	Losses colour in dilute acid	Poor to fair	Poor to fair	Sensitive
Lutein	Yellow	Excellent	Good	Good	Sensitive
Lycopene	Yellow to red	Good	Good	Fair	Fair
Paprika (Capsanthin)	Orange-red	Fair	Fair to good	Poor to fair	Sensitive

Source: Morley, 2014

the European Food Safety Authority reasserting that food irradiation is safe. It is still used for many products, such as herbs and spices, but is less common with finished products because of the labelling requirements.

Other techniques to minimise the use of additives such as heat processing (pasteurisation and ultra-heat treatment (UHT)) are often available but are known to cause thermal degradation, which may lead to non-desirable sensory impact. Hurdle technology is an approach which combines the inherent parameters within a product – pH, aW, etc., and external influences such as processing to create a series of stages and gates which, when combined and considered together, provide the traceable food safety assurance. This allows for an informed minimal usage of a preservation additive.

12.6.3 Antioxidants

Oxidation is a primary cause of food degradation, be it oxidative rancidity of fats, discolouration of meat products or enzymic browning of cut fruit. Each results in shelf life reduction. Amongst the commonly used antioxidants are butylated hydroxyanisole (BHA) and butylated hydroxytoluene (BHT), both linked to carcinogenic effects that are driving research for alternatives, such as plant-derived polyphenols, being considered as alternatives (Garcia-Lomillo and Gonzalez-SanJose, 2013). Packaging solutions have also contributed to successful prevention of oxidation through modified atmosphere packaging (MAP) and more recently skin packaging, both of which are designed to control or exclude the presence of oxygen. Other innovations, such as oxygen scavengers, can be included in bagged products, e.g. potato crisps, to provide a preferential sacrificial oxidation pathway, thus extending the life of the product. Such packaging options can provide an ideal solution to removal of additive antioxidants.

12.6.4 Sweeteners

There are many artificial sweeteners available, from the early saccharin, cyclamate, acesuflame K, aspartame and, more recently, sucralose, stevia and advantame, to name just a few. As each of these provides different sweetness profiles and relative intensities, they are often blended to provide a more rounded organoleptic perception. In general, these are considered with mixed feelings: on the one hand, they provide an effective alternative to sugar to help calorie reduction in the diet; on the other hand some, such as aspartame, have been linked to a range of health concerns. This consumer negativity around aspartame has led to the launch of products such as the sucralose-sweetened Yoplait Light™ product in June 2014, carrying an overt "free from aspartame" declaration. It has also fuelled the appetite for more natural image sweeteners such as stevia, which has seen much wider use globally since receiving EU approval in 2011 (Kumar and Guleria, 2012) and in the USA in 2008, despite having been widely used in China and South America for at least 40 years (Heyden, 2013). An interesting development in this arena is the introduction in 2014 of "Coke Life™" sweetened with a blend of sugar and stevia, providing a half-way house for the reduced calorie soft drink market, and building on the combined natural image of sugar and stevia over typical artificial options.

12.6.5 Emulsifiers, stabilisers, thickeners and gelling agents

This group of additives are essential in providing products with stability in some form, particularly with regard to quality parameters on shelf, whether in ensuring the product stays in its emulsion form, or preventing separation.

Emulsifiers attract some scrutiny in terms of label image. Soya and egg lecithin (E322) meet the recognised ingredient test, but other components: *mono-* and *di-*glycerides of fatty acids (E471) or polyglycerol polyricinoleate (PGPR) (E476) are somewhat confusing to the consumer. Emulsions such as chocolate and toffees would not exist in a stable form without them. Soya and egg derived lecithin each carry their allergenic warning and so, depending on the requirements of the product, provide their own labelling consideration.

Emulsions can be fragile, each emulsifier providing a very specific function depending on its degree of lipophilicity, and are often used in conjunction with other emulsifiers. Such doubling up may suggest the use of unnecessary ingredients, but the specificity of their function means that that this cannot always be avoided. Alternative emulsifying systems based on protein–polysaccharide conjugates, plant sterols and fats with naturally high levels of mono/diglycerides have been explored, but in broad terms have not been found to provide robust solutions (Clegg *et al.*, 2009), due to the complex nature of each emulsion system.

Thickeners are typically polysaccharides (starches) and in some cases overlap with the gelling agent category with the use of pectins or gums (vegetable or gelatine). These components benefit from their plant origin, or their description as flour on a label. Because of this, combined with a consumer's knowledge of the application of gels and thickeners in a domestic setting (gravy and jelly), their use is more widely accepted and understood.

Concerns around meat derivative safety throughout the 1990s, combined with an increase in demand for vegetarian products, have prompted a search for gelatine alternatives. However, due to the unique thermo-reversible nature of a gelatine gel, only selected polysaccharides have been identified as partial replacements (Marrs, 1999).

12.6.6 Others

The "other" category covers many different aspects of food additives, too numerous to fully consider here. However, within this group, modified starches are often subject to concern relating to the degree of processing they receive. Starches can be physically, chemically or enzymatically treated to deliver a broader or more specific degree of functionality and enable the starch to become more robust to the processing (Titoria *et al.*, 2011). However, with names such as "hydroxy propyl distarch phosphate," they are generally not accepted at the point of end use. Clean label starches are very much the focus of commercial development, and the successful launch of the Novation® starches (Ingredion), such as the patented Etenia™ and Eliane™ from Avebe, has seen functional starches that can be labelled as simply starch becoming available. Research continues in this important arena, since despite such positive technology they are yet to provide a full product range to replicate the functionality from cross-linked starches in pasting, gelation, viscosity, but also in terms of stability.

12.6.7 Fat

Fats in general suffer from the scrutiny and debate of the health agenda, and there is much debate regarding the relative issues surrounding saturated and unsaturated fats, with recent suggestions that saturated fat may not be as bad for human health as previously thought (Watson and McNeil, 2014).

The use and subsequent removal of hydrogenated oils provides an interesting test case and demonstration of a successful industry action to remove them from products. The underlying concern was, of course, in relation to the formation of "*trans*-fatty acids" during the partial hydrogenation process. Hydrogenation is the process of passing hydrogen through an oil to change its setting characteristics and so its functionality. During this process of saturation *cis*- and *trans*-fatty acids formed. The *trans*- form is generally not abundant in nature and was thought to contribute to greater degree to heart disease than the *cis*- form. Since this was a function of partial hydrogenation, full hydrogenation could have provided a solution. However, as the acceptability of labelling terms is dictated by the consumer and retailer, trust in the technology had been lost and use of hydrogenation to any degree was marginalised.

It also demonstrates the potential for swing in consumer opinion, as hydrogenation is the process that enabled margarine, once promoted as a healthier alternative to butter.

A further point to make within the context of label declarations is the need to meet the ethical demands of the public. Deforestation to create farm land to grow palm and soya, in particular, has driven the demand for the use of sustainable palm oil, which in turn has led to the formation of organisations such as the Roundtable for Sustainable Palm Oil (RSPO). Such ethical aspects must not be overlooked in delivering a label that the target market expects.

12.6.8 Sugars

It is a reasonable conjecture that the use of sugar will become increasingly scrutinised in the light of the recent establishment and activity by Action on Sugar. Sugar consumption as a whole is being linked to worrying increases in global obesity.

Sucrose, the di-saccharide, and its mono-saccharide components, glucose and fructose, are the most widely recognised sugars. High fructose corn syrup (HFCS) is increasingly treated with suspicion as concerns are raised about the body's ability to process fructose and studies linking it to increased cases of type 2 Diabetes (Chan, 2012). This renders HFCS out of scope for a clean label formula, despite fructose's common descriptor as "the fruit sugar." Sugar is considered by many to be over-processed, and with suggestions that it is added to excess, it is advised that the use of sugar is always in moderation, and where possible avoided, to uphold the integrity of the label.

This is, however, where the clean label paradox is perpetuated. Sugar performs a range of functions within a product, delivering sweetness, flavour, flavour Maillard reaction precursors (Osborn, 2013), texture, structure and preservative properties. To remove the sugar and incorporate the range of functionality that a single recognisable ingredient performs may prove challenging. Combinations of intense sweeteners and sugar alcohols (polyols) are common alternatives to sugar, since they have a reduced calorie profile. The labelling issue arises, since polyols are known to have laxative effects and advisory comments are required on-pack warning about the effects of excessive consumption. They are also less sweet than sucrose, and often require additional sweetness enhancement. However, as we have discussed, blends of sweeteners may be required to deliver the taste and sweetness profile. There is little dispute regarding excessive sugar consumption not being beneficial to long term health, but additional labelling requirements and replacement of a single, arguably natural, ingredient with three less identifiable alternatives may be detrimental to the healthy perception of the reformulated product.

12.6.9 Fibre

Fibre is increasingly being seen as having great potential for clean labelling. Dietary fibre, such as the naturally occurring polysaccharide, inulin, provides low calorific bulk to a product, and is also suggested to be beneficial to digestion and gut health through pre-biotic activity. It is also frequently used as a fat substitute, as the smooth texture makes it suitable for this function (Hall, 2012).

Such fibre is therefore ideal in its ability to satisfy a broad range of label needs, as well as the potential to provide functionality and structure to the product.

12.6.10 Protein

Amongst the ideals in formulating a label that meets consumer expectations is that of providing a recognisable nutritional balance. It therefore follows that protein is an essential element. Proteins are widely regarded as natural, as their origin can be assigned, whether animal or vegetable. They also have the benefit that, with the exception of gelatine, they are not allocated an "E" Number (Alting *et al.*, 2012). Proteins as whole can provide a range of functional benefits to a product, as their structure and behaviour can range from random to gluten and hydrocolloid-like and clustered.

With positive health messages being associated with protein (Chung *et al.*, 2014; Leidy *et al.*, 2014) and strong associations to its origin, it can deliver a positive labelling message provided it is able to deliver the functionality.

It is worth using protein as cautionary example of the need to maintain sight of the product context. The fundamental premise of clean label is that the ingredients are recognisable and not over-processed. Protein research is currently exploring a wide range of enzyme treatments that can enhance their functionality, from solubility to water-binding properties. It may even be fair to draw parallels to the exploitation of starch through chemical modification. This presents the likelihood that proteins could be employed in products that may seem out of context based on their functionality. For example, potato proteins in beverages may convey the image of over-processing. This would, however, be for the customer to feed back.

12.7 Future trends

Clean label presents a range of conflicts and issues, such as stability and shelf life; while consumers are welcoming of "natural" and "clean label" products, they are less forgiving of poor and dull colours, and in some cases seasonal variation. A report from Leatherhead Food Research (2011) explained the "balancing act" manufacturers must perform in adopting the use of "clean label" ingredients, as in every case a product must:

- Deliver a consistent product quality outcome
- Enable the manufacturer to remain competitive
- Be consistent with consumers' perceptions of naturalness

There are a number of key trends that remain strong throughout the global food and drink industry. Health and wellness in many ways is "business as usual" for the product developer. New products have to be made with low levels of fat, salt and sugar, which presents an array of challenges within an New Product Development

(NPD) environment; this can be of particular concern when set against the consumer demand for "clean label" and "natural" products. Additives are often an essential functional component of a food product, and without clear definitions of "natural" or "clean label" the consumer can easily be confused or misinformed, despite the best intentions of the food industry. While the concept of sugar, salt and fat removal may be deemed as ideal, especially when considering the overall drive for healthier foods, they are recognisable ingredients and the alternatives may not be in line with consumer expectations for that product. This creates a paradox between healthier product development and satisfying the need for "clean label." Resolution of this dichotomy can only come from transparency, consumer education and trust in the food and drink industry.

Further progress will be made through improving the technical performance and reducing the costs of production for natural ingredients to help maintain the price and quality positioning of their products. This goes hand-in-hand with increased investment and innovation to accelerate the widespread adoption of natural ingredients. However, this can only be achieved by educating consumers as to the use and function of certain ingredients, and more use of language that a consumer can understand in listing ingredients; this in turn will help enhance trust in the food industry.

To achieve long term success in "clean labelling" a globally accepted definition will be difficult, since the consumer and the industry perceptions and realities may be different. It is therefore probably more appropriate to define a philosophy of clean label that can inform a holistic reformulation strategy.

12.7.1 Clean label philosophy

To achieve a clean food product the objectives are:

- Minimise the range of ingredients used in order to simplify the label, and in doing so select necessary ingredients that can be labelled in a way that the consumer understands and recognises.
- With each ingredient, understand its broader function and consider how best to maximise it within a product.
- Where additives are a necessity, seek natural alternatives and convey its functionality so that the consumer understand its importance.
- On no account should food safety be compromised, but minimal processing techniques should be considered at every given opportunity.
- Engage with the consumer to ensure that they accept the ingredients used, understand their role and consider them to be appropriate for the product under consideration.
- Employ a sense of flexibility and discretion to allow informed and structured changes to products.
- Remember not all additives are bad and "natural" is not always good – even in the eyes of the consumer.

It is always important to revert to first principles, to understand your customers' motivations to be making such demands. It is easy to say that the premises are wrong, yet customers do hold the success of the product in their hands. One of the primary drivers of this trend was a lack of trust in the food and drink industry, and the feeling that there were hidden "evils" in food. A "clean label" philosophy should not therefore be one that excludes all additives, but one that ensures those used are necessary and in line with your consumers' expectations and needs; therefore, being open about the fact that a product needs and includes a preservative should be considered as "clean label." Transparent and honest labelling is essential to ensure this approach is trusted.

12.8 Sources of further information and advice

There are continuing developments in the clean label arena, the following being some of the more active and notable:

Cargill (www.cargill.com / http://truvia.co.uk/) Suppliers of Stevia
Diana Naturals (http://www.diana-food.com/) Suppliers of clean label proteins
GNT (www.gnt-group.com) Suppliers of natural colours
Ingredion (http://www.ingredion.com/) Suppliers of clean label starches
Overseal (www.overseal.com) Suppliers of natural colours
Purac (www.purac.com) Supplier's preservatives and antioxidants
Solanic (http://www.solanic.eu/Home.aspx) Suppliers of clean label proteins
Ulrick and Short Ltd. (http://www.ulrickandshort.com/) Suppliers of starch, flour, protein and fibre

Media channels such as Food Navigator.com, Food Manufacture (www.foodmanufacture.co.uk) provide useful and on-going commentary on product and ingredients development as well as consumer trends.

Leatherhead Food Research (www.leatherheadfood.com) provide reformulation expertise for clean label products, supported by global regulatory composition labelling expertise. Output from Leatherhead Food Research, such as Daily Food News, provides a useful aggregation and distillation of the changing landscape of food and beverage health, nutrition, labelling and trends.

References

Alting, A.C. and Van De Velde, F. (2012) Proteins as clean label ingredients foods and beverages, in Baines, D and Seal R (eds), *Natural Food Additives, Ingredients and Flavourings*, Woodhead Publishing Limited, Cambridge, 197–211.

Beeren, C. (2010) Consumer perceptions of clean label. *Food Com*, **14**:1.

Berry, D. (2011) Clarifying Clean Label, Food Product Design, http://www.foodproductdesign.com/articles/2011/05/clarifying-clean-labels.aspx# First accessed 27 June 2014.

Chan, A.L. (2012-11-27) "Is there a link between high fructose corn syrup and diabetes rates?" *Huffington Post*. First accessed 30 June 2014.

Chung, C., Degner, B. and McClements, D.J. (2014) Development of Reduced-calorie foods: Microparticulated whey proteins as fat mimetics in semi-solid food emulsions. *Food Research International*, **56**, February 2014, 136–145.

Clegg, S., Titoria, P.M. and Aragoneses, C. (2009) Alternatives to Chemical Emulsifiers, Leatherhead Food Research Forum Project Report No. RR941.

Delves-Broughton (2012) Natural antimicrobial's as additives and ingredients for the preservation of foods and beverages, in Baines D and Seal R (eds), *Natural Food Additives, Ingredients and Flavourings*, Woodhead Publishing Limited, Cambridge.

Diaz, J. (2013) How to clean up your label (part 1). The World of Food Ingredients. April/May 2013.

Food Standards Agency (2008) Criteria for the use of the terms fresh, pure, natural etc. in food labelling. Accessed on 14 May 2014: http://multimedia.food.gov.uk/multimedia/pdfs/markcritguidance.pdf.

Food Standards Agency. Current EU approved additives and their E Numbers Accessed on 24 June 2014: http://food.gov.uk/science/additives/enumberlist.

Garcia-Lomillo, J. and Gonzalez-SanJose, L. (2013) Plant Antioxidants as food additives. *Food Science and Technology*, **27**:3, 14–17.

Hall R.H. (2012) Applications of natural ingredients in savoury food products, in Baines D and Seal R (eds), *Natural Food Additives, Ingredients and Flavourings*, Woodhead Publishing Limited, Cambridge, 281–317.

Heyden, T. (2013) How did Stevia get mainstream? BBC news 4th June 2013; http://www.bbc.co.uk/news/magazine-22758059 accessed 27/06/14.

Kumar Yadav, S. and Guleria, P. (2012) Steviol glycosides from stevia: biosynthesis pathway review and their application in foods and medicine, *Critical Reviews in Food Science and Nutrition*, **52**:11, 988–998.

Leatherhead Food Research (2011) Clean Labelling – Consumer Perceptions and Implications for Food and Drink Manufacturers. Surrey: Leatherhead Food Research.

Leidy, H., Ortinau, L., Rains, T. and Maki, K. (2014) Acute effects of higher protein, sausage and egg-based convenience breakfast meals on postprandial glucose homeostasis in healthy, premenopausal women, April 2014. *The FASEB Journal*, **28**:1, Supplement 381.6.

Marrs, M. (1999) Alternatives to Gelatine: A Review, Leatherhead Food Research Association, Scientific and Technical Survey, 191.

McCann, D., Barrett, A., Cooper, A., Crumpler, D., Dalen, L., Grimshaw, K., Kitchin, E., Lok, E., Porteous, L., Prince, E., Sonuga-Barke, E., Warner., J.O. and Stevenson, J. (2007) Food additives and hyperactive behaviour in 3-year-old and 8/9-year-old children in the community: a randomised, double-blinded, placebo-controlled trial. *Lancet*, **370**:9598, 1560–1567.

Morley, W., Terlicka, K. and Osborn, S. (2013) Healthy solutions. *Food and Drink Technology*, **12**:8, 16–17.

Morley, W. (2014) Flavours and Colours, Confectionery Production July/August 2014. **80**: 4.

Osborn, S. (2012) Special report: Definition of a Flavouring. Beverage Innovations. December 2012, 26.

Osborn, S. (2013) Milking the benefit of Maillard, Confectionery Production, June 2013, **79**:4.

Osborn, S. (2014) Clean Label: Paradoxically simple?,The World of Food Ingredients. June 2014, 10–15.

Sahi, S. (2012) Applications of natural ingredients in baked goods, in Baines D and Seal R (eds), *Natural Food Additives, Ingredients and Flavourings*, Woodhead Publishing Limited, Cambridge, 318–331.

Scott, J. (2010) Defining Clean Label. International Food Ingredients No. 3 June–July 2010, United Business Media. Available at http://www.cleanlabelinsights.com/Documents/IFI%20Article%20CLEAN%20LABEL%20DEFINITION%202010.pdf. First accessed 28 June 2014.

Titoria, P., Arrongeness, C. and Groves, K. (2011) Alternatives to Chemically-modified Starches. Leatherhead Food Research Forum Project Report No. 947.

Teoh, A. (2012) Natural Colours update, *Beverage Innovations*. October 2012.

Watson, E. and McNeil, G. (2014) FoodNavigator Interview, Time for a rethink on Sat. fat? 30-Jun-2014; http://www.foodnavigator-usa.com/Suppliers2/Time-for-a-rethink-on-saturated-fat. First accessed 30 June 2014.

Wyers, R. and Tremellen, N. (2013) *Innova Makert Insights Top Trends, The World of Food Ingredients*. December 2013, 11–12.

Index

Printed in the United States
By Bookmasters